Sheffield Hallam University
Learning and IT Services
Adsetts Centre City Campus
Sheffield S1 1WS

AF743688

101 942 157 6

SHEFFIELD HALLAM UNIVERSITY
LEARNING CENTRE
WITHDRAWN FROM STOCK

ON

SHEFFIELD HALLAM UN
LEARNING CENTR
WITHDRAWN FROM ST

Aviation and Climate Change

Routledge Studies in Physical Geography and Environment

This series provides a platform for books that break new ground in the understanding of the physical environment. Individual titles focus on developments within the main sub-disciplines of physical geography and explore the physical characteristics of regions and countries.

1. Environment Issues in the Mediterranean
J. Wainwright and J. B. Thornes

2. The Environmental History of the World
Humankind's Changing Role in the Community of Life
J. Donald Hughes

3. History and Climate Change
A Eurocentric Perspective
Neville Brown

4. Cities and Climate Change
Harriet Bulkeley and Michele Betsill

5. Little Ice Ages
Ancient and Modern
Jean M. Grove

6. Himalayan Perceptions
Environmental Change and Well-Being of Mountain Peoples
Jack D. Ives

7. The Natural History of the Earth
Debating Long-Term Change in the Geosphere and Biosphere
Richard John Huggett

8. Aviation and Climate Change
Lessons for European Policy
Alice Bows with Kevin Anderson and Paul Upham

Aviation and Climate Change

Lessons for European Policy

Alice Bows
with Kevin Anderson
and Paul Upham

Routledge
Taylor & Francis Group
New York London

First published 2009
by Routledge
270 Madison Ave, New York, NY 10016

Simultaneously published in the UK
by Routledge
2 Park Square, Milton Park, Abingdon, Oxon OX14 4RN

Routledge is an imprint of the Taylor & Francis Group, an informa business

© 2009 Alice Bows, Kevin Anderson and Paul Upham

Typeset in Sabon by IBT Global.

All rights reserved. No part of this book may be reprinted or reproduced or utilised in any form or by any electronic, mechanical, or other means, now known or hereafter invented, including photocopying and recording, or in any information storage or retrieval system, without permission in writing from the publishers.

Trademark Notice: Product or corporate names may be trademarks or registered trademarks, and are used only for identification and explanation without intent to infringe.

Library of Congress Cataloging in Publication Data

Bows, Alice, 1974–
Aviation and climate change : lessons for European policy / by Alice Bows, Kevin Anderson and Paul Upham. -- 1st ed.
p. cm. -- (Routledge studies in physical geography and environment ; 8)
Includes bibliographical references and index.
1. Aeronautics--Environmental aspects. 2. Aircraft exhaust emissions--Europe. 3. Aeronautics--Energy conservation--Europe. 4. Aircraft exhaust emissions--Environmental aspects. 5. Climatic changes. 6. Environmental policy--Europe. I. Anderson, Kevin L., 1962– II. Upham, Paul, 1966– III. Title.

TD195.A27B69 2008
629.134'35--dc22 2008008903

ISBN10: 0-415-39705-7 (hbk)
ISBN10: 0-203-89189-9 (ebk)

ISBN13: 978-0-415-39705-6 (hbk)
ISBN13: 978-0-203-89189-6 (ebk)

SHEFFIELD HALLAM UNIVERSITY
WL
363.73874
BO
ADSETTS LEARNING CENTRE

Contents

Figures

Tables

Acknowledgments

We would like to acknowledge the *Tyndall Centre for Climate Change Research's* financial support for the research that underpins this book. The Centre is itself funded by the Engineering and Physical Sciences Research Council (EPSRC), Social Sciences Research Council (ERSC) and the Natural Environment Research Council (NERC). Additional funding sources supporting elements of the research presented include Friends of the Earth and the Joule Centre (funded by the North West Development Agency).

We would also like to thank the School of Mechanical, Aerospace and Civil Engineering within the Faculty of Engineering and Physical Sciences and Manchester Business School, both at the University of Manchester, for providing support for Tyndall Manchester.

Individuals who we would like to acknowledge who have made useful contributions to the research or who have given their support and help during the book's creation include: Sarah Mander, Sally Randles, Holly Waterman, Chris Jones, Richard Starkey, Mercedes Bleda, Simon Shackley, Carly Mclachlan, Philip Boucher, John Broderick, Anthony Footitt, Mike Hulme, Andrew Watkinson, Asher Minns, Susan Stubbs, Mike Childs and Richard Dyer.

Finally, we would like to thank aviation industry and policy representatives who have taken part in the research over the previous four years.

1 Flying into Heavy Weather

INTRODUCTION

The UK has a buoyant aviation industry with a larger number of passengers passing through its airports than any other European nation. Yet far from being mature, UK aviation continues to benefit from high rates of growth year-on-year. The emerging importance of the low-cost business model has only served to fuel this trend, with air fares falling rapidly as low-cost airlines take advantage of the liberalised market. Although, in the UK, only 50% of the adult population flew in the previous twelve months, this proportion has grown in recent years.[1] Furthermore, many individuals are choosing to fly more frequently, no longer viewing air travel as a luxury, but more often a means to an end. The fact that the UK's aviation market is arguably more mature than many, suggests this global industry is set to continue to grow rapidly as more nations industrialise. Indeed evidence for this is clear in China, where passenger numbers have been growing on average at 15% per year since 1990.[2]

A successful aviation industry is often viewed as a core element of economic prosperity, generating jobs and investment, and as a key to globalisation. However, contemporary industrialised societies are facing a contradiction through their simultaneous desire for mobility powered by fossil fuels, and a high-quality natural environment. Climate change is proving an intractable problem. Whilst for many sectors, reducing emissions to levels commensurate with avoiding 'dangerous climate change' appears to be viable, it is considerably more challenging for aviation given current rates of growth and limited annual fuel-efficiency improvements.

This book aims to explore the barriers and opportunities relating to absolute emission reduction in the aviation sector within the context of a transition to a low-carbon economy. By drawing together Tyndall Centre research, both the scale and urgency of the climate change challenge are addressed. Moreover, reasons for the recent spotlight on the aviation industry are explained by setting aviation within the context of the other fossil-fuel consuming sectors. The research findings presented here address many of the issues on the global, EU and national scales. In some cases, research is presented in both an older and more up-to-date form, to illustrate the surrounding dynamic political and scientific environment. The book therefore provides a useful reference for understanding the evolution of the aviation and climate change debate during the first seven years of the twenty-first century. However, given the current pace of developments in relation to public acceptance of climate change and growing engagement with the climate change issue from the aviation industry (at least in Europe), the picture may well be quite different in only a few years.

For the remainder of this introductory chapter, each section provides an overview of the subsequent chapters. Although the chapters in isolation provide useful reference material, the particular sequence was chosen to build up the arguments for urgent climate change mitigation policies and emphasise the significance of rapidly growing sectors such as aviation. The book culminates with a comparative analysis of aviation both in the context of a low-carbon economy and in relation to the other energy-consuming sectors.

AVIATION'S PAST, PRESENT AND FUTURE

Following this introductory chapter, the book begins with an overview of recent aviation development from the perspective of trends in global passenger numbers, passenger-kilometres travelled and corresponding estimates of the carbon dioxide (CO_2) emissions from aviation. This analysis is then broken down to consider, in more detail, the EU and UK's aviation industries. Apart from short-term impacts due to external shocks, for example the events of September 11, 2001, the industry has been growing at rates well above national or global GDP since the 1970s, and currently shows no sign of abatement. Although the aviation industry has improved its fuel efficiency over the same period, CO_2 emissions have also grown due to the rates of growth exceeding the improvements in fuel efficiency. It is, therefore, only when the industry manages to reduce the amount of fuel consumed per passenger-km by the same rate as the growth in passenger-km, that CO_2 emissions will be stabilised, unless of course, lower-carbon fuels begin to penetrate the fleet.

Traditionally, the aviation industry has appeared resilient to the various shocks affecting individuals' propensities to fly; wars, terrorist attacks, health scares, for example. Whereas very often these previous shocks occurred in isolation, terrorism, high oil prices and a new public and policy concern for the environment are happening in unison, at least within the UK. The resilience of the industry is therefore being tested, but as yet it is too early to judge the outcome.

The future of aviation is a concern of many internal and external to the industry, from providing adequate airport infrastructure and manufacturing an appropriate quality and quantity of aircraft, through to assessing the sector's environmental impacts. In general, the aviation industry tends to use economic forecasting for future visioning, with models premised on economic relationships, such as the historical link between GDP growth and energy consumption. As a consequence, these tools may be obsolete when considering radically different futures where historical links are substantially changed. Breaking the relationship between economic growth and CO_2 emissions and/or energy consumption is essential if a growing world economy is to make the transition to a low-carbon economy. Tools such as scenario methodologies will therefore play an increasingly dominant role in future visioning, if all industries, including aviation, are to understand their roles in carbon mitigation.

For aviation in particular, CO_2 is not the only emission of concern. Aircraft emit NO_X, soot and water vapour which can all lead to climate change. Confusion over how to account for these additional emissions continues to blight policy formation. By attempting to quantify the *historic* impacts of these other emissions in relation to aviation's CO_2 emissions, scientists have unintentionally confused policymakers and those producing carbon calculators into thinking it is possible to multiply CO_2 emissions from a flight by a simple 'uplift factor' to estimate total climate impact. Given the danger of influencing decision makers into well-meant but ill-informed policy measures, such as altering flight altitude at the expense of additional fuel consumption, the analysis within this book focuses on CO_2 alone. This is not to suggest that these other emissions should be ignored, but rather they should be considered separately. Clearly it is important for any comparative analysis to state what emissions are or are not included, and to describe the assumptions underpinning the emissions with which aviation is being compared. Therefore, when comparing the emissions contribution from aviation with the associated contribution from other sectors, it is important to firstly consider them on a like-with-like basis. Secondly, comparing aviation emissions within the context of long-term growing global emissions is not helpful in assessing aviation's contribution during the transition to a low-carbon world. These points are covered in some detail in Chapter 2.

CLIMATE CHANGE AND CUMULATIVE EMISSIONS

Over the course of the research conducted at the Tyndall Centre, new insights have been gained regarding the key areas of concern in relation to climate change science and policy. The importance of understanding the role of cumulative emissions in shaping the future climate is one such area to be discussed in Chapter 3. As CO_2 remains in the atmosphere for over one hundred years, each year's emissions broadly add to those emitted during the previous century. Consequently, it is the magnitude of cumulative emissions over time that is a dominant factor in influencing the resultant atmospheric concentration of CO_2 in the atmosphere, and this CO_2 concentration can then be linked with a global temperature response. As a result, when deriving climate policy, it is the cumulative emissions that are crucial, as opposed to a certain percentage reduction by a particular year. By using a cumulative budget approach to inform global, EU and UK carbon mitigation, it is clear that avoiding 'dangerous climate change' becomes much more challenging than when basing it on long-term reduction targets.

Two areas of uncertainty within the cumulative budget approach relate to *Climate Sensitivity* and *Carbon-Cycle Feedbacks*. Climate sensitivity is the long-term temperature response to a doubling of the atmospheric concentration of CO_2 in the atmosphere. In recent years, this value has increased from a mid-range value of 2°C to one of 3°C.[3] Consequently, the global atmospheric CO_2 concentration to be aimed for, to remain within a reasonable chance of not exceeding the 2°C warming associated with 'dangerous climate change', becomes significantly lower. The question for policymakers is, how does this impact on national emission pathways? Adding to this is the further uncertainty relating to *Carbon-Cycle Feedbacks*. The amount of CO_2 that remains in the atmosphere is dependent on the activity of the carbon cycle; how much CO_2 is absorbed by the earth and oceans. It is the remaining unabsorbed CO_2 that gives rise to a particular CO_2 concentration level, such as 450 parts per million by volume (ppmv). However, recent research indicates that as the global temperature increases, feedback loops may lead to a reduction in the absorption of CO_2 by the land and oceans.[4] As a result, humans can release less CO_2 into the atmosphere for a particular CO_2 concentration level than previously thought. Coupled together, scientific advances in understanding the climate sensitivity and carbon-cycle feedbacks reduce the total amount of CO_2 that can be emitted for a given temperature increase. By presenting both pre- and post-climate sensitivity and carbon-cycle feedback improvement results for the EU and UK, the consequences for climate mitigation policy become clear (see Chapter 3).

OPPORTUNITIES FOR AVIATION

The mitigation challenge spelt out in Chapter 3 emphasises the need for all regions, nations and sectors to play their role. The aviation sector too will need to consider all of the possible opportunities for it to improve its energy efficiency, reduce its energy consumption and modify its reliance on high-carbon fossil fuels. A summary of these potential opportunities are presented in Chapter 4. Although it appears that the industry, certainly within the UK and EU, is now engaged with the climate change issue, there are many technological and operational barriers to change, not least in relation to the long time lag involved in complete global fleet renewal. Nevertheless, if all of the opportunities presented in Chapter 4 could be exploited, the industry would certainly be going a long way towards addressing its growing climate impact.

Technological opportunities for the aviation sector include modifications to engine and airframe designs. A variety of much more efficient airframes are currently in operation, but primarily within the military. Transferring the technology to the civil aviation sector presents a significant challenge. Technology is clearly not the only constraint; safety concerns, public perceptions and operational adjustments for radically different aircraft must all be overcome. Given that current civil aircraft have typical lifetimes of thirty to forty years, radical technological change within the timescale necessary is limited by the current regulatory framework; technological solutions are not simply about science and engineering!

Low-carbon fuel sources, including low-carbon electricity, hydrogen, renewables and biofuels, are now being explored by almost all sectors of the energy system. Aviation however, is less amenable to such fuel-switching. Hydrogen aircraft will likely require a completely new airframe design, coupled with a worldwide hydrogen fuel infrastructure. Biofuels, on the other hand, may offer a potential low-carbon aviation fuel, although sustainability issues will need to be taken into account and are likely to prove constraining.

From an operational perspective, there appear to be a plethora of opportunities for change. Indeed, some of the low-cost airlines are already managing to relatively reduce the amount of fuel consumed per passenger through the use of a newer fleet, higher load factors and increased seat densities. Making changes to minimise fuel wasted from indirect flight routing could be managed through the incorporation of an improved European air traffic control system. Congested airports and airspace may also be improved, again resulting in a reduction in fuel per passenger-km. However, freeing up the skies by improving efficiency could lead to an absolute increase in fuel consumption, if such a move boosts growth. Addressing this rebound effect is essential in carbon mitigation policies.

CLIMATE AND AVIATION POLICY

To better understand the overarching political and regulatory framework within which aviation is tackling climate change, Chapter 5 presents an overview of the current state of play. Interestingly, it is this chapter that has required constant amendment due to the very fast pace of policy development relating to aviation and climate change.

International aviation and shipping emissions were omitted from inclusion within the national emission inventories required to be submitted as part of the Kyoto Protocol. Instead, the United Nations Framework Convention on Climate Change (UNFCCC) requested the International Civil Aviation Organisation (ICAO) to put measures in place to address aviation's impact on the climate. Unfortunately, despite ICAO requesting the International Panel on Climate Change (IPCC) to produce its Special Report on

Aviation,[5] the EU Commission recently concluded ICAO had done little to address the growing impact of aviation on the climate. As such, the EU has taken it upon itself to incorporate not only EU flights, but all departures and arrivals from and to the EU within its own emissions trading scheme (EU ETS). The effectiveness or otherwise of the scheme with aviation included is uncertain. Nevertheless, the EU is at least exhibiting the necessary leadership on the aviation–climate issue.

The UK Government has itself supported including aviation within the EU ETS and, in addition, recently increased air passenger duty to apparently address aviation's growing climate impact, a particularly unpopular move with the industry. On the other side of the coin, the UK Government is continuing to support its 2003 plans laid out in its Aviation White Paper for considerable expansion of the UK's airports. The conflict between the 2003 Aviation White Paper's aims and the same government's 2003 Energy White Paper's goal of reducing CO_2 by 60% by 2050 are presented graphically in Chapter 5. Whilst the UK Government continually restates its commitment to the global 2°C temperature threshold target, it clearly continues to view the aviation sector as a special case. Unfortunately, given its goal is a global mean temperature, omitting certain sectors from its own analysis renders its targets meaningless.[6] Only when this omission is addressed and quantified by Government, can it claim to base its targets on a firm scientific evidence-base.

COMPARATIVE ASSESSMENT

The surrounding policy framework, outlined in Chapters 3 and 5, along with the opportunities for the aviation sector to reduce its carbon intensity in the context of current and future trends and drivers summarised in Chapters 2 and 4, are brought together using scenario analysis within Chapter 6. By developing aviation emission scenarios at both the EU and UK levels and comparing them with the target emission pathways in line with 2°C, EU and UK policymakers are able to make an assessment of the importance of the aviation sector compared with their overall carbon budget.

Within Chapter 6, both old and new analyses are presented to illustrate the very rapidly moving policy and scientific frameworks within which this issue is being tackled. The older analyses point to a requirement for the industry to moderate growth rates and to make fuel efficiency (or carbon intensity) improvements as a matter of some urgency, in order to remain within the carbon budget available. The more up-to-date analyses consider aviation emission scenarios within a cumulative carbon budget commensurate with the 450 ppmv CO_2 concentration and push the aviation industry to deliver much more radical carbon intensity improvements than are currently being considered. As such, although aviation's emissions consume a significant amount of the carbon budget by 2050, and in some cases the entire allowance, it is concluded that an acceleration in fuel efficiency and low-carbon fuel developments, coupled with more moderate growth, could allow the industry to remain within a carbon budget based on 2°C. However, even then, all of the other economic sectors of the EU or UK will be required to compensate for the aviation sector's, at best, marginal emission reductions from 2000 levels by 2050.

AVIATION IN THE WIDER ENERGY CONTEXT

The aviation sector has been the focus of academics, policymakers and nongovernmental organisations (NGOs) in recent years. Some within the aviation industry feel this focus is unreasonable given aviation's share of emissions today compared with other sectors, such as private car transport. Indeed car transport contributes a higher proportion of emissions than aviation at present, but given climate change analysis is concerned with future mitigation and adaptation issues, it is those sectors likely to be contributing significantly to climate change in the future that are of key interest at present. It is an inescapable fact that the aviation industry is growing rapidly in both industrialised and industrialising nations, with relatively few options for step changes in reducing its carbon intensity. This is not, however, the case for many other transport modes or indeed the household, industrial and service sectors. In Chapter 7, scenarios for the UK's energy system illustrate a range of aviation's contributions to future CO_2 emissions compared with the emissions from the other economic sectors. Again, both old and new analyses are presented, the older being in line with the UK Government's 60% carbon-reduction target, and the newer commensurate with the 450 ppmv cumulative emission pathway.

If all sectors were to continue along current trends in relation to energy efficiency improvements and growth in activity, the UK's carbon emissions would likely be somewhat higher in the future than they are currently. However, within a UK playing its fair role in achieving the 2°C target, the future emission budget is severely limited. By exploring where it is easiest to reduce energy consumption through energy efficiency measures and behavioural change, in addition to reducing the carbon intensity of energy supply through low-carbon electricity and alternative low-carbon fuels for transport and heat, new visions for the UK's energy system are presented. Some sectors are more constrained than others in terms of the opportunities available to them. For example, if the most efficient four-person car currently on the market became the mean efficient car for the entire UK fleet, radical improvements in the energy efficiency of the car sector could be achieved in a relatively short space of time. Similarly, if by 2050, the electricity grid became dominated by renewable technologies, nuclear power and coal-fired power stations with carbon capture and storage, household CO_2 emissions from electricity could be minimal.

The very long lifetime and design lifetime of aircraft, allied with incremental improvements in fuel efficiency and the slow pace of international regulations governing the industry, serve to reduce the viable opportunities available to aviation to reduce its CO_2 emissions, particularly when compared with many other sectors. Consequently, it may be necessary for other sectors to compensate for aviation emissions if growth in the sector is to continue. By 2050, aviation will likely become a dominant carbon emitting sector, even if the sector pushes the boundaries of technology and operational efficiency.

The choices are therefore uncomfortable for EU nations; either governments remain committed to the 2°C threshold, and accept that their greenhouse gas emission targets must include all sectors, or they must revisit the target, and accept more significant climate change impacts. The former will require stringent action to curb aviation growth to ensure efficiency improvement rates are similar to passenger-km growth rates; the latter necessitates informing those working on adaptation to prepare for climate change impacts commensurate with average temperature rises of well in excess of the 2°C target.

2 Aviation
Past, Present and Future

INTRODUCTION

Tackling climate change requires an understanding of all sources of greenhouse gases as well as wider climate warming and cooling emissions. The majority of the sources within industrialised nations are energy consuming sectors such as transport, households and industry. To most effectively and rapidly reduce emissions, it is likely that some of those sectors will be able, and therefore required, to reduce their emissions more than others, depending on available mitigation options. However, without a thorough understanding of each sector's climate impact in the past, at present and in the future, it is not possible to develop successful mitigation policies. Furthermore, given the very short timescales over which action must be taken to tackle climate change (see Chapter 3), mitigation options considered commensurate with avoiding 'dangerous climate change' are increasingly limited. This chapter will present the historical growth of the aviation industry, the current state of the industry and future aviation forecasts, including a discussion of an alternative method of future visioning. Finally, it will provide an overview of this industry's current contribution to climate change and discuss some of the confusion associated with future emission projections and scenarios.

THE PAST

In the first section of this chapter, historical trends in passenger and freight traffic as well as CO_2 emissions are presented in the context of the globe, EU and UK. The trends are broken down into these spatial scales in line with this book's emphasis on national and regional (e.g. EU) climate policy. Climate change targets are often set on national and regional scales, and therefore, despite the international nature of aviation, it is not only practical but appropriate to discuss aviation's contribution within this context.

The Global Scale

During the twentieth century, the rate of worldwide energy use increased ninefold,[1] with the most rapid growth in demand arising from electricity use and transport. Although the transport debate has often revolved around personal car use, interest in the aviation industry has risen rapidly up the political agenda, particularly in European nations. Since 1960, global air passenger traffic (expressed as revenue passenger-kms or RPK[2]) has increased by nearly 9% per year—2.4 times the growth rate of global average GDP.[3] Similarly, air cargo transport has grown by over fifty times, reaching around 200 billion revenue freight-tonne-kms or FTKs in 2006. By 2006, there were over 18,000 commercial aircraft in service, operated by some 1300 airlines, passing through approximately 1200 airports and flying over 4 billion RPKs.[4]

Historically, energy consumption has been closely linked to changes in global GDP. Similarly, the aviation industry has grown as nations have industrialised. However, rates of growth within national aviation industries have typically exceeded national GDP growth, and this continues to be the case in many industrialised nations, with the UK being no exception. A range of indicators can be taken as measures of growth within the aviation sector, with each having differing levels of importance for aviation stakeholders. For example, airports may take a particular interest in the growth in passenger numbers in order to understand and maximise retail opportunities, whereas manufacturers may be more interested in the growth in the number of flights or RPKs. From the perspective of climate change, the most important parameter of interest is the growth in greenhouse gas emissions and other emissions affecting the radiative properties of the atmosphere. In terms of long-lasting impact, the most important greenhouse gas released by aircraft is CO_2, which is directly proportional to the aviation fuel consumed.

The relationship between the change in the number of RPKs, passengers or flights compared with the change in fuel consumption, and hence CO_2 emissions, depends on a variety of factors. For example, an increase in passenger numbers may be brought about through increasing the number of people on board the same aircraft for the same flight. In this case, growth in the passenger numbers and RPKs will result in virtually no growth in the CO_2 emissions (although a small increase will occur due to the slightly higher weight being carried by the aircraft). There will be no growth in flights in this case. On the other hand, the same number of passengers may be transferred from A to B in a larger, less fuel efficient aircraft, resulting in no growth in passenger numbers, RPKs or flights, but an increase in CO_2. Another variation is when a larger number of passengers are transferred in one larger plane, rather than two smaller planes. In this case, the passenger numbers increases, RPKs may remain the same (one journey rather than two), flight numbers decrease, and CO_2 emissions may also decrease, depending on the aircraft's efficiency. Finally, a new version of the same aircraft may be 20% more fuel efficient per RPK than its older counterpart, but if the growth in the number of RPKs is greater than 20%, then there will be no reduction in the CO_2 emissions. It is therefore important to consider the range of factors influencing CO_2 to understand aircraft emission mitigation.

To gain a better understanding of how the industry has been growing in the past, a number of indicators are presented here. Figure 2.1 illustrates how global passenger numbers have been growing since 1970. Over the past twenty years, global air passenger numbers have increased at an average of around 5.2% annually. The data presented in Figure 2.2 illustrates how global RPKs have been growing since 1970. Over the past twenty years, global RPKs have increased at around 4.8% per year—a slightly lower growth rate than for passenger numbers. The corresponding growth rate for freight is 7.6% per year, and aircraft movements, 4%. So, how does CO_2 emission growth compare?

Over the past twenty years, CO_2 emissions from flights have increased at 2.7% per year (Figure 2.3). As passenger, rather than freight traffic, dominates the aviation industry, this suggests fuel efficiency improvements over the previous twenty years have been around 2% per year. In their special report on aviation, the IPCC suggest aviation's fuel efficiency improves at around 1 to 2% per year.[6] As long as the growth in both RPK and FTK

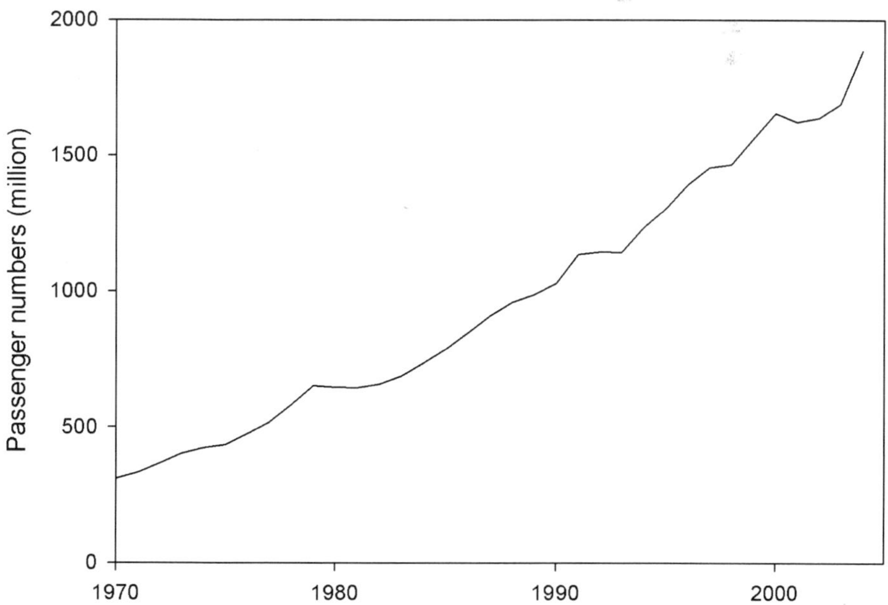

Figure 2.1 Global passenger number growth since 1970 (World Development Indicators April 2006 edition[5]).

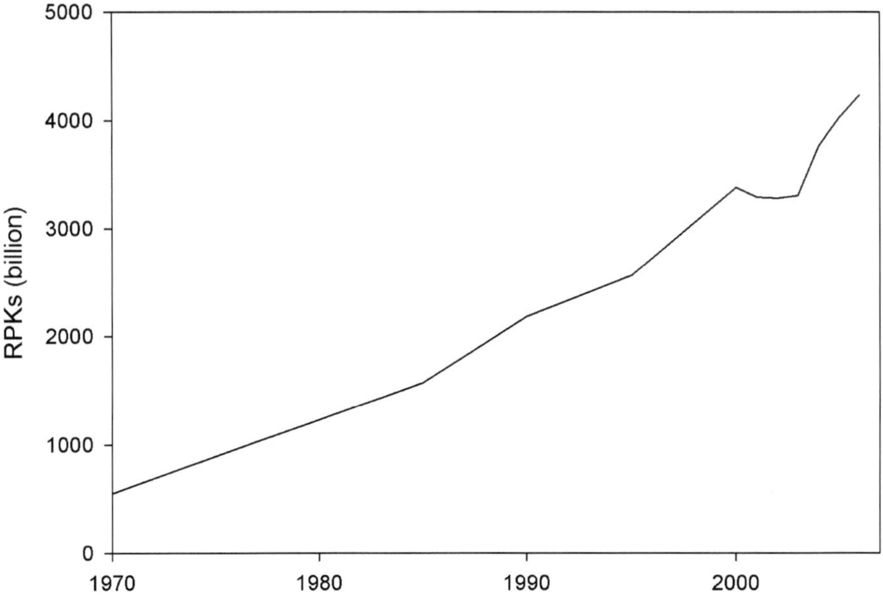

Figure 2.2 Historical development of revenue passenger-km (Boeing and Airbus market forecast data).

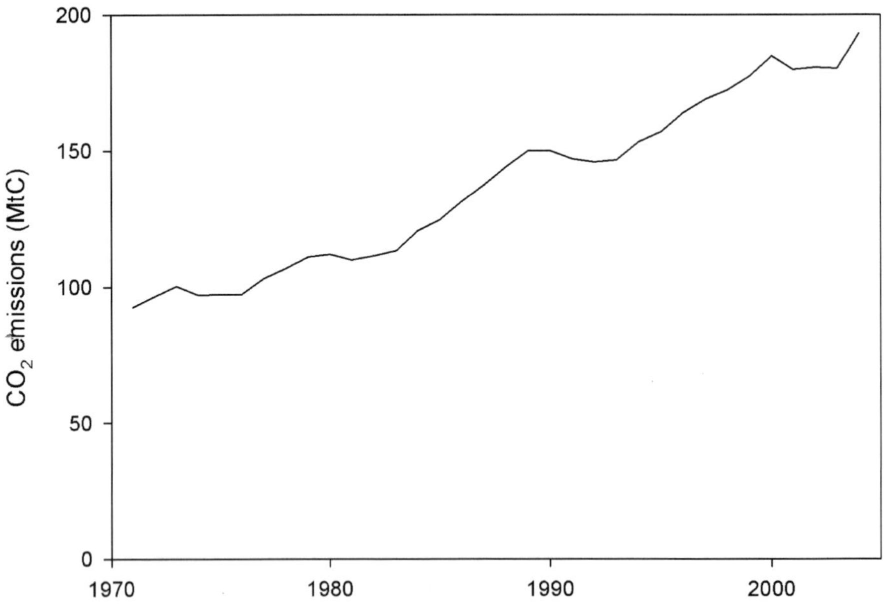

Figure 2.3 Global CO_2 emissions from aircraft flights (World Development Indicators, April 2006 edition).

exceeds the improvements in fuel efficiency (per either RPK or FTK), emissions from this sector will continue to grow. This concern forms a core element of this book.

The EU Scale

At the time of writing, the EU had recently incorporated Romania and Bulgaria, resulting in twenty-seven nations being fully within the EU. However, as the majority of the research was conducted prior to these two nations' inclusion, the discussion presented here will focus only on the EU25 nations.

There has been a step change in recent years in relation to the opportunities for airlines to operate on routes across the EU. Prior to liberalisation in the 1990s, national or 'legacy' carriers dominated the market. Since then, young airlines such as the low-cost carriers have generated new opportunities for passengers to travel not only to more destinations, but for significantly reduced fares. It has been argued by some within the industry that the low-cost carriers have not increased the growth rate within the EU, but merely substituted for the markets previously dominated by legacy airlines. Whatever the reasons behind recent growth in EU aviation, flying is growing in popularity not only in the richer of the EU nations, such as the UK, but particularly within the less industrialised European nations such as Poland. As a consequence, the EU's carriers are experiencing continued high growth.

As some nations joined the Union later than others, historical data for the EU25 is not always readily available. However, published reports indicate passenger numbers within the EU15 nations increased at 5.3% per year between 1993 and 2002 and in the ten accession nations, passenger numbers increased at a similar level between 1995 and 2000. More recently, EU25 passenger numbers in 2005 exceeded 700 million, with an 8.5% increase on the previous year's figures.[7] This level of growth illustrates a resurgence of the industry following the events of September 11, 2001. Although data for passenger numbers for the whole of the EU25 is readily available only for the years 2004 to 2005, typical rates of growth can be gleaned from individual nations' statistics. Figure 2.4 illustrates passenger growth in three contrasting exemplar nations: UK, Poland and Spain.

None of the nations illustrated in Figure 2.4 are at the extremes of the industry's passenger growth range; the UK's passenger numbers increased by an annual average of 7% between 1993 and 2005, and at a similar level between 2003 and 2005. Spain's respective figures were 10% and 9%, whereas Poland's grew at an annual average rate of 21% between 2003 and 2005.[8] Although it might be expected that passenger growth would have a close relationship with CO_2 emissions growth, this link is not evident from the available data. Figure 2.5 illustrates this point.

Nations within the EU are required by the Kyoto agreement to submit national emission inventories on a yearly basis to the United Nations

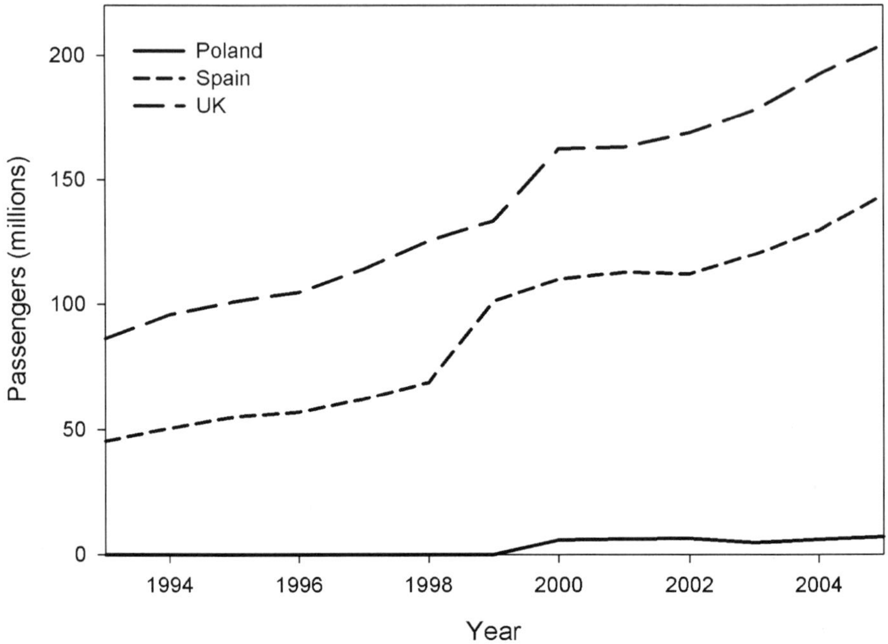

Figure 2.4 Passenger number data from three exemplar nations within the EU25 (EU Commission data[7]).

Framework Convention on Climate Change (UNFCCC). However, it is not a requirement that nations collect CO_2 data for international flights for their national inventories; instead international 'bunker' emissions generated by aviation and shipping are submitted as a memo to the UNFCCC. Domestic flight emissions are included within national inventories.

Comparing the passenger and emission growth trends for the UK and Spain, based on the UNFCCC submissions, shows an unclear relationship between passenger growth and corresponding CO_2 trends. In the UK, passenger numbers[10] increased by around 7% per year between 1993 and 2005, the corresponding CO_2 emissions grew at around 6% on average. For Spain, despite a 10% per annum increase in passenger numbers between 1993 and 2005, its CO_2 emissions only increased at 6% per year. The fact that emissions are growing relatively more rapidly per passenger for the UK than Spain could be attributable to a disproportionately higher growth in long-haul travel compared with medium- and short-haul. On the other hand, it could be an indication of deficiencies in the CO_2 or passenger number estimates. Interestingly, despite a fast-growing aviation industry within Poland, emissions were reported as decreasing from 1993 levels by 2005. Given it is understood that only incremental improvements to aircraft fuel efficiency over the last two decades have been made, this would appear to suggest inconsistent datasets, indicating significant discrepancies

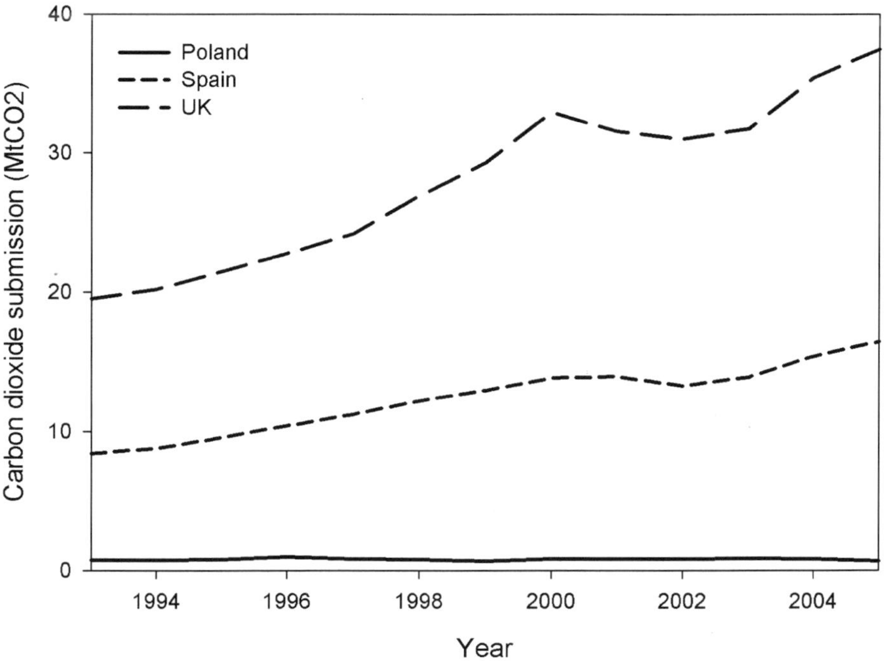

Figure 2.5 CO_2 emissions from three exemplar nations within the EU25 (UNFCCC submissions[9] for domestic and international aviation bunkers).

between data collection standards and robustness across the various EU nations. This situation requires remedying as a matter of urgency if our understanding of the link between growth trends and emissions is to be improved.

Turning now to the emissions generated by the EU25 nations' aviation industries, on aggregate, the latest data, as submitted to the UNFCCC in 2007, is presented in Figure 2.6. It is clear from Figure 2.6 that emissions from international flights dominate. CO_2 emissions from domestic flights have increased at an average of 2.5% per year since 1990 according to this dataset. The corresponding figure for international flights is 4.5%. However, the events of September 11, 2001 had a marked impact on the growth rate of the aviation sector, as illustrated in the data between 2001 and 2003. If the period between 1990 and 2000 is considered, domestic aviation's annually averaged CO_2 growth was 3.2%, with international air travel at 5.6%. This period also incorporated the first Gulf War, which understandably impacted on the industry. From 2003 to 2004, and 2004 to 2005, the total amount of CO_2 from the EU25's aviation industries increased by 7% and 6% respectively. Between 2004 and 2005, passenger numbers increased by 8.5%. Therefore, current rates of growth in CO_2 emissions are similar to those produced by the industry between the gulf war in 1993 and the period affected by the events of September 11.

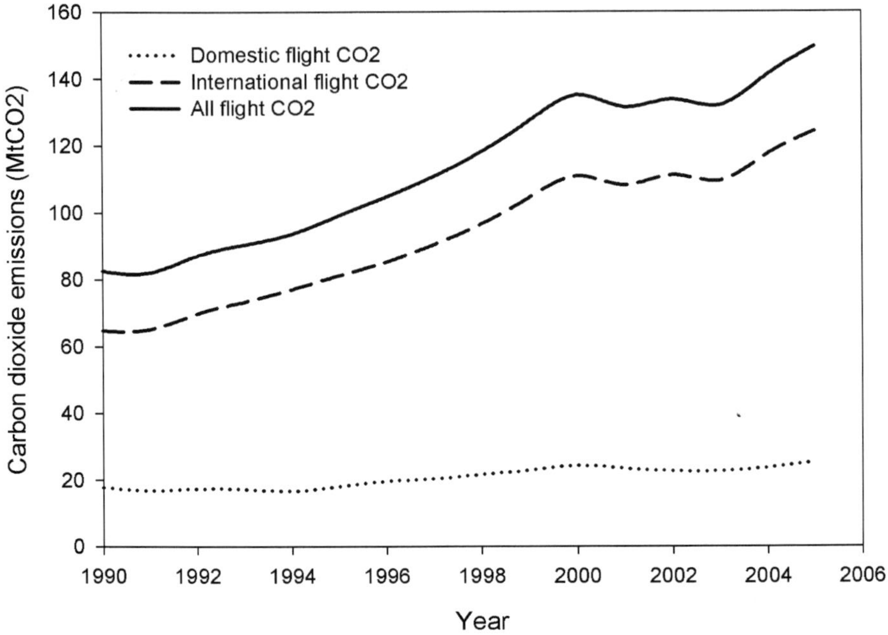

Figure 2.6 CO_2 emissions from the EU25's aviation sectors, from data submitted to the UNFCCC in 2007. The data incorporates an estimate for Greece and Malta in 2005 due to an absence of data at the time of writing. Although not all of the EU25 were in the EU from 1990, all of the nations have been included in the totals from the outset.

The UK Scale

The UK tops the European chart in terms of terminal and transfer passenger numbers. Its largest airports act as key hubs for many trans-Atlantic and transcontinental flights. Furthermore, the UK has been central to the low-cost airline revolution, with Easyjet and Ryanair having large bases at several of the UK's airports. Despite the UK's aviation industry being considered relatively mature, growth in terms of passenger numbers has averaged over 5.5% per year between 1990 and 2005, including the downturns due to the first Gulf War and the events of September 11, 2001 (data from CAA[11]). Between 1993 and 2000, excluding these events, passenger numbers grew at 7% per year. Activity indicators for the UK's aviation industry are presented in Figure 2.7.

Prior to the events of September 11, 2001, average annual growth rates between 1990 and 2000 were as follows: Terminal passengers—6% per year; Air Traffic Movements—4% per year; Seat km used by UK airlines—8% per year; Kerosene consumption—5% per year; CO_2 emissions—7% per year. Following the events of September 11, there has been a resurgence in UK aviation, with elevated rates of growth seen between 2003 and 2004

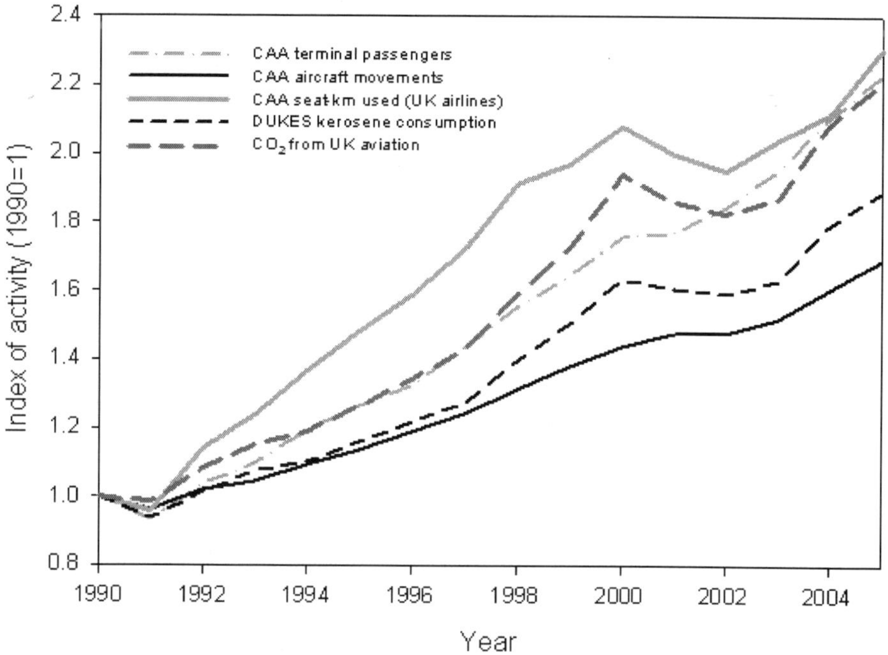

Figure 2.7 Recent UK aviation trends: traffic, energy use and CO_2 emissions.

of 8% in terms of passenger numbers[11] and a corresponding increase in fuel consumption for that period of 10%.[12] 2004–2005 saw fuel consumption and hence CO_2, growing at a somewhat lower rate of 5%. The UK's aviation industry appears to exhibit significant resilience to major incidents, however, whether or not this resilience will continue given the current combination of security fears, high oil prices, airport employee strikes and rising environmental concern amongst consumers, remains to be seen.

THE FUTURE

Notwithstanding periodic shocks, demand for fast and reliable global air transport currently looks set to continue under prevailing market conditions. For industries to plan for the future, forecasts and/or future visioning is required. From Tyndall's aviation industry interviews and workshops, it is clear that economic forecasting is the principal tool for future planning within the aviation sector. However, in a number of other sectors, including energy and transport, alternative methods such as scenario construction are frequently employed. Scenarios provide an opportunity to paint both a quantitative and a qualitative picture of the future without relying solely on potentially constraining historical trend data. Although economic forecasting may be considered viable in periods of stable economic development, the reality is that external shocks to the system are always possible. Furthermore, the current climate change debate, requiring urgent and significant action, arguably renders conventional economic forecasting inappropriate. For example, if economic growth is to continue, and climate change to be addressed successfully, then a break in the relationship between economic growth and aviation kerosene consumption will be necessary. However, at present, this relationship is incorporated mathematically within economic forecasting models. Only with the use of alternative techniques like scenarios can policy and decision makers, both within and external to the industry, explore new, alternative and sustainable visions for aviation's future.

Uncertainties and Shocks

Within this section, some of the economic forecasts published by the aviation industry and the UK Government are presented. Later in the book, scenarios will be developed taking an alternative view of how the industry might develop in the future, in the context of nations striving to reduce their climate change impacts. First, however, issues relating to dealing with uncertainties and shocks are discussed.

Although there are a number of forecasts and predictions available for aviation at a variety of spatial and temporal scales, the future is inherently uncertain, and any one of the predictions can prove to be incorrect. Each forecast is generated with a specific purpose in mind. For example, Airbus forecast growth to plan the number and types of aircraft that they need to manufacture. The UK's Department for Transport (DfT), on the other hand, may be attempting to predict and provide for the necessary airport infrastructure to meet projected demand.

The emergence of the low-cost airlines, particularly in Europe, has thrown an additional level of unpredictability into the mix. Their entry into or exit from particular routes has the potential to substantially change traffic flows, positively or negatively. Due to the tighter margins within the low-cost model, where fuel costs are a much higher proportion of their overall costs, low-cost traffic is likely to be more susceptible to influence by external factors than is traffic carried by the more traditional, or so-called, 'legacy' airlines. This is not to say, however, the low-cost airline sector itself is a riskier prospect. One piece of research[13] lists the aspects of the sector—in a strengths, weaknesses, opportunities and threats assessment framework—that contribute to forecasting difficulties through inherent instabilities in the factors involved. Examples of each include: strength—fast turnaround time and high utilisation rates; weakness—less flexibility in terms of operation failure; opportunity—predicted increase in business passengers due to increasingly constrained business travel budgets; threats—increase in fuel taxes.

The Global Scale

The aviation industry produces forecasts of both future passenger numbers and flights for business strategies, infrastructure planning, hardware purchase and environmental concerns. Boeing and Airbus, the two main aircraft manufacturers, produce forecasts for a period of twenty years into the future, some of the details of which, along with forecasts from Rolls-Royce, are summarised in the following.

Airbus Forecasts

Airbus predicts global passenger traffic will grow at, on average, 4.8% per year between 2006 and 2025, with the world's airline fleet doubling over this period. This world average incorporates an average growth in RPK of some 7.2% per year in China. Such high levels of growth are reflected in the increased share of world air traffic for the Asia-Pacific region for 2025 from some 26% in 2006 to 32% in 2025. In 2005, 4 trillion RPK were travelled globally compared with a predicted 10.5 trillion by 2025.

The Airbus forecasts use projections from the Global Insight Forecasting Group and take into account anticipated economic growth, oil prices and the export/import history of countries and commodities. Regional and structural changes expected to influence the market are also included, for example the growth potential for low-cost carriers. The pace of liberalisation of markets to and from industrialising countries, as well as environmental and congestion constraints, are also considered.

Boeing Forecasts

Recent forecasts from Boeing are similar to those issued by Airbus. Again, they predict world passenger growth will, following the downturn after the events of September 11, 2001, return to a rate of 4.5% per year, RPKs at around 5% per year and cargo at 6.1% per year[14] between 2006 and 2026. Figure 2.8.

Rolls-Royce Forecasts

The engine manufacturer Rolls-Royce produces yearly forecasts for aircraft deliveries, engine markets and traffic forecasts. In 2007 Rolls-Royce[15] forecast that the in-service fleet of airliners is expected to double over the next twenty years. The fleet is not expected to increase as fast as traffic, due to improvements in aircraft utilisation and load factors, as well as a "modest" increase in average aircraft size. They predict continued strong long-term growth in all major segments of the commercial aircraft and jet engine market, anticipating demand for 132,000 engines over the coming twenty years. The engine manufacturer expects to deliver engines to some 30,389 business

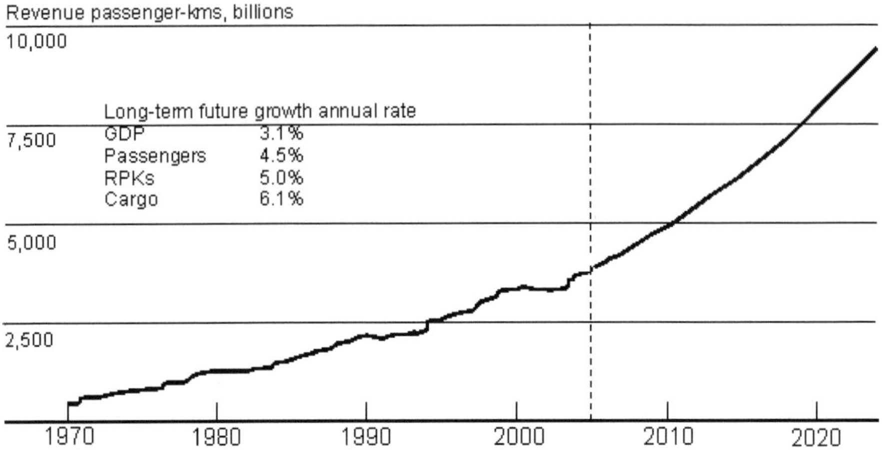

Figure 2.8 Boeing forecast for global air passenger traffic.[14]

jets, 6,558 regional jets of 30–90 seats and 23,315 mainline aircraft. Of the latter, demand for aircraft of 130–180 seats is predicted to exceed the second largest category of 300 and 350 seat aircraft by a factor of four. Markets within Asia, both short-haul and intercontinental, are expected to drive much of this growth. However, the more mature markets in Europe and North America are expected to require over 6,000 new airliners to replace older aircraft in today's fleet. Rolls-Royce state that as well as anticipated demand for passenger airliners and cargo aircraft, a "huge" market exists for business jets. They observe that aircraft and engine makers have experienced strong sales in this segment since 2003, with growing demand outside of North America, the traditional home to around 80% of the global fleet. In their traffic forecasts, Rolls-Royce anticipate a global average annual growth rate in RPK of 4.9% over 2006–2026. Asia-Pacific has the strongest expected growth (6.6%) and North America the lowest (3.5%). Europe is expected to average 4.2% growth in RPK.

The EU Scale

For more specific data relating to Europe, the European Organisation for the Safety of Air Navigation, EUROCONTROL, produce short-, medium- and long-term forecasts for flights,[16] summarising predicted increases in the number of flights per year with growth disaggregated into particular regional flows. EUROCONTROL forecast traffic growth for individual EU states and produce scenarios of possible futures. Rather than produce forecasts in terms of passenger numbers or RPK, the data is given in terms of total aircraft movements. Over the medium term, flights within EUROCONTROL's Statistical Reference Area are predicted to increase by 3.4% between 2006 and 2013. For their long-term forecasts, EUROCONTROL use four different scenarios to generate a range of annual increases in flights. Their latest publications[16] produce growth rates of between 2.7% to 3.7% per year between 2006 and 2025 compared with an average growth of 3.8% per year between 1975 and 2001. They anticipate that the number of annual flights will increase at a lower rate in the longer-term future than over the coming five or so years, due to airport capacity constraints.

Airbus breaks down its global market forecasts into the different world regions, and predicts that for airlines domiciled in Europe growth[17] will be at around 4.6% per year between 2006 and 2025, with a higher rate of growth of 5.0% per year between 2006 and 2015. This forecast takes into account the mounting importance of low-cost carriers in Europe. In the Boeing forecasts for the EU, air traffic for the region's carriers is expected to be on the increase at 4.2% per year, somewhat lower than the previous twenty-year annual growth rate of 5.7%, although the rise of low-cost carriers is predicted to continue generating new growth in Europe. European markets have completed their first decade of liberalisation, which has rapidly stimulated air travel demand. Lower fares and point-to-point services to many secondary and select hub airports have proved popular with air travellers.

In summary, despite being a relatively mature industry, growth within the EU's aviation sector is predicted to continue at relatively high rates until the mid 2020s. Although emissions from the aviation sector are not referred to within the market forecasts discussed here, the limited opportunities for step changes in improvements to fuel efficiency[18] will lead to corresponding greenhouse gas emissions from aviation increasing at around 3 to 4% per year.

The UK Scale

In 2003, the UK Government published its Aviation White Paper,[19] including forecasts anticipating a near trebling of air passengers by 2030. Demand in 2006, according to CAA statistics[11], was in the region of 235 million air passengers, while the midpoint forecast presented in the White Paper for national demand for 2010 is 275 million, 2020 is 400 million and 2030 is 500 million passengers per year. This would result in annual average passenger growth of around 4% per year between 2010 and 2020 and 2% between 2020 and 2030. Regarding airfreight, UK demand doubled between 1989 and 1999 and is forecast to grow even more rapidly over the coming few years. The DfT forecasts show freight traffic in the South East increasing from around 2 million tonnes per year currently to 6 to 8 million tonnes by 2030.

In addition to the passenger number forecasts, the Aviation White Paper included forecasts for CO_2 emissions between 2000 and 2050, based on their passenger number projections. In total, they produced three emission growth forecasts within accompanying documentation;[20] one low growth ('best'), one medium growth ('central') and one high growth ('worst'). The 'worst case' forecast assumes limited fuel efficiency improvements, limited fleet renewal and no economic instruments. This forecast was based on the 'high capacity' case developed within another of the supporting documents, on economic instruments[21] with additional runways in the South East of England, as well as, so-called, unconstrained capacity in the regions.[22] The 'central case' figures are also based on the 'high capacity' scenario, but incorporate fleet fuel efficiency improvements envisaged by the IPCC[6] and by the Advisory Council for Aeronautics Research in Europe (ACARE). Finally, the 'best case' estimate incorporates the use of economic instruments to produce an additional 10% fleet fuel efficiency saving from 2020 onwards, with a 5% fleet fuel efficiency saving in 2010. These forecasts are for both domestic and international aviation and include freight transport.[23] The results are detailed in Table 2.1.

The data for these forecasts begin in the year 2000, but estimates for 2005 are now available. Furthermore, the methodology used to calculate the data has been revised recently and historical figures reevaluated. Comparing the published emission estimates for 2000 to 2005 shown in Table 2.2 with the UK Government's forecasts illustrates firstly, the starting figure for 2000 has now been estimated at 0.6 million tonnes of CO_2 ($MtCO_2$) higher. Secondly, that the figure for 2005 has already reached the Government's 'best' case prediction for 2010 and emissions would only be allowed to grow at 2% per year between 2005 and 2010 if they are to remain within the upper bound of the Worst Case estimate by 2010.

Therefore, the very high growth rates experienced by the aviation industry during the past few years have already rendered the UK Government's forecasts out-of-date, even when the events of September 11, 2001, are

Table 2.1 UK Government Aviation Forecasts for CO_2 Emissions

UK Government (DEFRA) CO_2 Emission Forecasts for Domestic Plus International Aviation In Million Tonnes of Carbon (MtC)			
Year	*Worst Case*	*Central Case*	*Best Case*
2000	8.8	8.8	8.8
2010	11.4	10.8	10.3
2020	16.5	14.9	13.4
2030	20.9	17.7	15.9
2040	25.1	18.2	16.4
2050	29.1	17.4	15.7

Table 2.2 Aviation CO_2 Emissions Data Submitted to the UNFCCC

CO_2 Emission Estimates for the UK's Aviation Industry (MtC)			
Year	*International*	*Domestic*	*Total*
2000	8.3	0.6	8.9
2001	8.1	0.7	8.8
2002	8.0	0.6	8.6
2003	8.2	0.7	8.9
2004	9.1	0.7	9.8
2005	9.7	0.6	10.3

factored in. Such poor forecasting, even for the short-term, seriously hampers the development of emission scenarios and hence meaningful carbon mitigation policies.

THE WIDER CONTEXT

Within this chapter, historical aviation trends have been presented in relation to indicators such as RPKs, passengers, flights and emissions for the global, EU and national scales. In addition, a few aviation industry, EUROCONTROL and government forecasts for both RPKs, passenger numbers, and in some cases, CO_2 emissions have been highlighted. By considering possible future aviation emissions, in the context of those emissions from other sectors of the world economy, it is possible to make an assessment of the relative contribution of the aviation sector to climate change. However, there are two distinct areas of concern in relation to how aviation's current and future emissions are presented that have, in the past, led to considerable debate and confusion. The first relates to the non-CO_2 emissions released by the aviation sector and the issue of radiative forcing, the second relates to predictions, forecasts or scenarios of future emissions for all non-aviation sectors. These issues will now be dealt with in turn.

Non-CO_2 Emissions and Radiative Forcing

When considering aviation's impact in relation to other sectors, the reader must understand the context within which aviation is being presented. As there is more certainty in relation to the climate impact and scientific understanding of CO_2 emissions than the other emissions released by aircraft, some studies compare the future emissions of CO_2 alone from aviation with the CO_2 alone from other sectors. Given the very long lifetime of CO_2 in the atmosphere compared with other non-CO_2 emissions from aviation, and taking account of the policy context within which climate change is being tackled, this is considered here to be the most appropriate method when considering future emission scenarios. On the other hand, some studies choose to make an estimate of the total amount of *radiative forcing* from the aviation sector in relation to all other sectors. This approach includes, in addition to CO_2, the non-CO_2 emissions from aviation, and consequently, in developing future policy, it is then necessary to consider changes in these other emissions as well as the CO_2 changes.

Whilst aircraft emit significant quantities of CO_2 into the atmosphere, their contribution to the level of warming is considerably greater than that associated with the CO_2 alone. This additional warming primarily arises from the altitude at which aircraft typically operate and consequently where their emissions occur. The science underpinning this additional warming is particularly complicated and a full understanding remains some way off. Nevertheless, current understanding does permit the principal components of the additional climate impacts to be identified and in some cases, quantified. Table 2.3 summarises the additional components arising from aircraft emissions at typical cruise altitudes.

Table 2.3 Emissions from Aviation and Temperature Impacts at the Earth's Surface

Emission Type	*Role*	*Principal Effect at Earth's Surface*
Carbon dioxide	Greenhouse gas	Warming
Water vapour	Greenhouse gas	Warming
	Contrail formation	Warming
Nitrous oxides	Forms ozone (greenhouse gas)	Warming
	Depletes methane (greenhouse gas)	Cooling
Sulphur oxides and sulphuric acid	Reflects sunlight	Cooling
	Contrail formation	Warming
	Increased cirrus cloud cover	Warming
Soot	Reflects sunlight	Warming
	Contrail formation	Warming
	Increased cirrus cloud cover	Warming

In addition to the uncertainty surrounding the science and subsequent quantification of some of these additional factors, there is considerable debate as to how they can best be compared with the warming attributable to CO_2 emissions. The simple numerically defensible approach is to compare them in terms of their instantaneous warming; generally referred to as radiative forcing. Radiative forcing is essentially a measure of the instantaneous change in the balance of solar radiation entering the atmosphere with the terrestrial and reflected solar radiation leaving the atmosphere. However, the additional impacts arising from aircraft emissions act over very different time scales; for example, whilst vapour trails (contrails) and cirrus clouds will have radiative forcing impacts over a few hours to a few days, and over a relatively small spatial area, the effect of nitrous oxide emissions on methane will impact radiative forcing from shortly after their release through to ten or more years.

Within the IPCC's *Special Report on Aviation*,[6] a somewhat modified version of the radiative forcing approach was adopted. In essence, the IPCC provisionally proposed that if the radiative forcing of all the additional factors were to be averaged across the globe and over the period affected by the emissions, the net warming from aviation emissions would be between two to four times that of the CO_2 alone. This figure, which excludes an additional warming impact of contrail-induced cirrus clouds, became known as the 'uplift factor'. In recent research, Sausen et al.[24] suggest that it is likely the figure (again excluding cirrus clouds) is closer to two times that of the CO_2 alone.[25] Despite the discussion of this factor within the IPCC, many scientists continue to express considerable concern over the appropriateness of a factor that includes the very short-lived warming impacts of vapour trails; all the more so as the vapour trails have a localised warming impact only, whereas CO_2 and, in many respects, the chemical influence of nitrous oxide, are well mixed within the global atmosphere.

An additional objection to the attention given to this 'uplift factor' is the misunderstanding of how it can be used. When the original calculation was made, it was done so to assess the historical impact of aviation emissions in relation to both other sectors and the impact of each emission source, for example, CO_2 compared with contrails. Therefore, it can be stated that over the past *y* years, NO_x and contrails from aviation have contributed *X* times the amount of warming than emissions of CO_2 alone, for example. It can *not* be used to say either of the following: a) NO_x and contrails emissions released by a particular aircraft are *X* times larger than the CO_2 emitted, or b) NO_x and contrails will, in some year in the future, contribute *X* times more warming than CO_2 alone. The quote[26] from a recent meeting of the Partnership for Air Transportation Noise and Emissions Reduction clearly spells out this concern:

> IPCC (1999) calculated the change in radiative forcing due to cumulated emissions from the historical aircraft fleet. It was a way to address

> the question of how the forcing will be different if no aircraft have been operating or a specific emission is omitted. It was NEVER intended that those radiative forcing numbers should be used directly in policy considerations. The convention of calculating RF in this way, as an absolute change in concentrations over a given period of time, is a "backward looking" approach that is not a suitable metric for comparing the impact of emissions (as compared to concentration changes) during some specified period on the subsequent climate change, because of the large contrasts in the residence time of the different contributors to RF (as discussed earlier in this report). IPCC (1999) also introduced the concept of the radiative forcing index (RFI), which is simply the ratio of the total forcing at a given time to the forcing from carbon dioxide at the same time. Unfortunately, the RFI has been misapplied in some quarters (as discussed by Forster et al. 2006) as a way of crudely accounting for the future non-CO_2 climate change impacts of aviation, by simply multiplying the CO_2 emission scenarios by the RFI. This example points to the need for scientific oversight of applications of climate metrics in policy considerations.

Given the remaining scientific uncertainty as to the mechanisms and quantification of the warming associated with vapour trails and cirrus clouds, allied with the time scale and mixing issues discussed earlier, the 'uplift factor' approach is inappropriate for developing and analysing projections, forecasts and/or scenarios of the future impact of the aviation industry. Moreover, in responding to the challenge of aviation's climate change impacts, the measures and policies considered here explicitly avoid approaches based on a single 'catch-all' metric. The more well-informed of the aviation industry, a well as those working on the climate science and associated analyses, are aware of both the additional climate change impacts of aviation and the complexities and uncertainties of the interactions between the different emissions. Ignorance of what many believe to be the scientific niceties surrounding 'uplift' has the potential to lead to undesirable consequences of well-meant but ill-informed policy. For example, altering the altitude at which aircraft fly would bring about a net reduction in the formation of contrails and cirrus clouds (with a consequent reduction in net warming over a few hours to days), but perhaps at the cost of increasing fuel burn, and hence CO_2 emissions (with a net increase in warming over the coming century).[27]

One final point relating to the use of the 'uplift factor' is an argument relating to comparing sectors on a like-with-like basis. In no other sector is the use of an 'uplift' (or 'downlift', as it could be in the case of shipping) factor commonly used. There is a study currently underway to quantify the radiative forcing effects of the non-greenhouse gas emissions from transport modes,[28] with results pending. However, comparing aviation with any other sector without including their additional emissions appears somewhat unreasonable. Only when the same emissions from one sector are compared with those from another, can the comparison be reasonably justified. For these reasons, any analysis presented within this book comparing future scenarios for aviation with emissions from other sectors, focuses on CO_2 alone.

The Global Future—Non-Aviation Sectors

In 2001, the IPCC devised a number of global scenarios for greenhouse gas emissions and published them within the *Special Report on Emission Scenarios* (SRES).[29] Each scenario made particular socio-economic assumptions in relation to, for example, GDP growth, land use, population, etc., driving, for example, energy consumption, agriculture, energy supply technologies and consequently emissions. A number of these 'SRES' scenarios result in a future likely to lead to very significant global temperature increases—well in excess of the 2°C commonly referred to within EU and UK policy documentation. In addition, some scenarios present futures where globally, greenhouse gas emissions have been stabilised, and according to the scientific knowledge at the time, produce global temperatures rising by not more than 2°C above preindustrial levels. In the next chapter, the significance of temperatures exceeding 2°C will be discussed. However, there are two points of importance to be stressed:

Firstly, if, for example, aviation emissions are being compared with one of the high emission growth scenarios, then, depending on the assumptions made in relation to global aviation growth, aviation's overall contribution compared with other sectors may be projected to be small. However, the reader must bear in mind that the world view presented in all but the B1, B2 and A1T scenarios is one subject to significant 'dangerous climate change', or with temperature rises in excess of 2°C by the end of the decade. A second point of note is one to be discussed in Chapter 3; due to an improved scientific understanding of the relationship between CO_2 emissions and temperature since 2001, emission reductions in excess of those presented in the 2001 IPCC scenarios will be required if the world is to avoid exceeding the 2°C threshold. Clearly, the more other sectors reduce their emissions in order to avoid 'dangerous climate change', the larger the contribution will be from any sector that neglects to reduce emissions. Within Chapter 6, emissions from aviation will be considered in relation to the other sectors within both the EU's and the UK's economies. This analysis will assume the EU and UK have been successful in addressing climate change, and therefore the scenarios are more closely aligned with the SRES B1, B2 and A1T scenarios than the other SRES scenarios.

SUMMARY

This chapter has examined elements of the aviation industry's history and presented a selection of future forecasts on the global, European and UK

Table 2.4 Aviation's Recent Estimated Contribution to Energy-related and Industrial CO_2 Emissions (CDIAC; UNFCCC Submissions; IEA CO_2 Data)

2004/5 Figures	*Total CO_2 Emissions from Energy and Industrial Processes (MtC)*	*Total CO_2 Emissions from Aviation (MtC)*	*Percentage Attributable to Aviation*
Global (2004)[31]	7910	193	2.4%
EU25 (2005)[32]	1152	41	3.4%
UK (2005) (UNFCCC 2006)	163	10	6.3%

scales. Difficulties in visioning the future were highlighted and confusion relating to the context within which aviation's impact on the climate is often debated was discussed. Given some of the existing misunderstandings, Table 2.4 summarises the 2005 contribution of the aviation sector to energy-related CO_2 emissions. Note the figures are presented in terms of CO_2 alone, and do not include the non-CO_2 impacts of aviation. The totals with which they are compared are a) the global total for energy CO_2 emissions taken from the Carbon Dioxide Information Analysis Centre (CDIAC),[30] b) the EU's total energy CO_2 emissions as submitted to the UNFCCC and finally c) the UK's energy CO_2 emissions as submitted to the UNFCCC. Globally, aviation in 2004–2005 is estimated to have been responsible for approximately 2.4% of total CO_2 emissions from fossil fuel combustion. For the majority of industrialised nations the figure is considerably higher, with, for example, aviation representing over 6% of UK CO_2 emissions.

Although in many respects the UK aviation industry is more mature than many of its EU counterparts, this relatively mature industry continues to experience very high levels of annual growth. Some of this growth is driven by expanding markets between the EU and Asia, but intra-EU flights continue to prove popular year after year. Consequently, not only are the current emissions from aviation already significant, they are subject to growth rates substantially greater either than most nations' economic growth (e.g., aviation growth in China is typically twice GDP growth) or indeed global GDP growth. It is therefore a continuation of the high levels of aviation growth across both industrialised and industrialising nations, which is of chief concern within the climate change debate.

3 Climate Change and Cumulative Emissions

INTRODUCTION

Within the UK and the EU there is broad political agreement that we should make our fair contribution to avoiding dangerous climate change. In the absence of any explicit global consensus on an appropriate metric for delineating dangerous from acceptable climate change, European leaders have frequently reiterated their commitment by suggesting the EU takes the lead internationally to "ensure that global average temperature increases do not exceed pre-industrial levels by more than 2°C".[1] The UK 2006 Climate Change Programme,[2] the 2006 Energy Review[3] and the 2003 Energy White Paper[4] all serve to underline the UK Government's commitment to the importance of the 2°C threshold. The link between a global temperature threshold, such as 2°C, to a national or regional carbon emission pathway is partially value-driven (e.g., through choosing a method of apportioning global emissions to regions or nations) and is further complicated by a range of scientific uncertainties (e.g., climate sensitivity). Nevertheless, understanding the link is useful when considering appropriate national and regional carbon-reduction strategies. This chapter will focus on understanding the consequences for climate policy of establishing a target of a 2°C global temperature rise for the global, EU and UK scales. It will then focus on emission apportionment for the purpose of forming regional and national climate change targets, with results presented for both an old and new interpretation of apportioned emission pathways for the EU and UK.

GLOBAL CLIMATE CHANGE

As global emissions of greenhouse gases increase, the radiative properties of the Earth's atmosphere are modified through an enhanced greenhouse effect. The consequences in terms of climate and weather-related impacts however, are not straightforward or necessarily predictable. This is due to the dynamic and interactive nature of the atmosphere in relation to radiation, atmospheric fluid dynamics, biogeochemical interactions and the various feedback mechanisms at work throughout this complex system.

The latest IPCC report[5] links its global greenhouse gas emission scenarios taken from the *Special Report on Emission Scenarios* (SRES)[6] with "best estimate" temperature increases of between 0.6 and 4°C.[7] However, recently published research illustrates that, due in no small part to the ongoing very rapid growth in energy consumption within China and India, the rate of growth of CO_2 emissions between 2000 and 2005 exceeds even the 'highest' (4°C) IPCC scenario.[8] As cumulative emissions of CO_2 have more influence on the ultimate temperature rise than do emission pathways[9] due to the long lifetime of CO_2 in the atmosphere, rising global levels of CO_2 today vastly reduce the options for emission mitigation over the coming years and decades. As such, global emission trends impact directly on regional and national carbon mitigation policy and therefore regional and national-scale policy must be considered in the evolving global context if regions such as the EU, and nations like the UK, continue to aim for a 2°C threshold.

Cumulative Budgets

A very simple example illustrating the importance of cumulative emissions is presented in Figure 3.1 and Figure 3.2. In Figure 3.1, the CO_2 emissions compiled by the CDIAC[10] from fossil fuels and industrial emissions, in addition to a low estimate of deforestation emissions based on figures within the Global Commons Institute's Contraction & Convergence model, are presented for the years 2000–2004. Estimates in line with data provided by Greg Marland from CDIAC (private communication) provide data for 2005 and 2006. Within the latest IPCC report, a range of cumulative CO_2 emission values are provided for given CO_2 stabilisation levels. For example, if CO_2 concentrations were to stabilise at 450 ppmv, the cumulative emission range given for the twenty-first century is between 1376 [375] and 2202 [600] $GtCO_2$ [GtC].[11] The mid-range value is given as 1798 [490] $GtCO_2$ [GtC]. Different levels of CO_2 stabilisation can be associated with probabilities of exceeding particular temperature thresholds. For example, stabilisation at 550 ppmv CO_2 equivalent (i.e., including the other greenhouse gases) offers a mean value of 28% of not exceeding 2°C and around a 46% chance of not exceeding 3°C. If CO_2 alone is being considered, then by estimating the contribution of other greenhouse gases as approximately 50 ppmv, and assuming limited net aerosol and non-greenhouse gas emission cooling and warming contributions in the future, 550 ppmv CO_2 offers a 88% chance of exceeding 2°C and a 75% chance of exceeding 3°C. Similarly, the 450 ppmv CO_2-alone level relating to the cumulative budgets described here, offers a 30% chance of not exceeding the 2°C threshold.[12] For reference, atmospheric concentration at the time of writing was around 384 ppmv CO_2 alone. The CO_2 equivalent value is not so readily available due to arguments relating to the emissions that should rightfully be included—i.e., should it

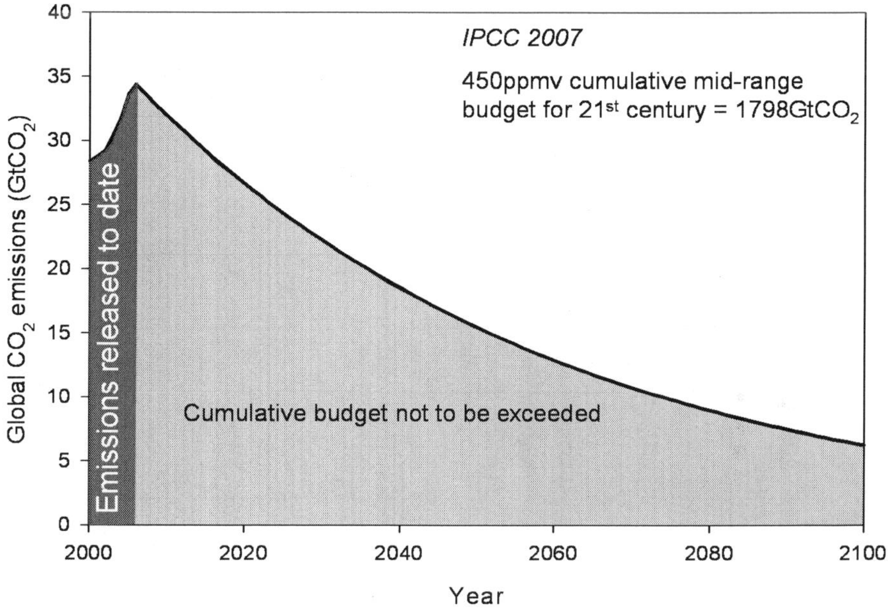

Figure 3.1 Estimated CO_2 emissions released until 2004 from energy, fossil fuels and deforestation from CDIAC, CO_2 estimates for 2004–2006 and an emission-reduction curve in line with cumulative CO_2 emissions not exceeding the IPCC's twenty-first-century budget for stabilising CO_2 concentrations at 450 ppmv.

include just the basket of six greenhouse gases or in addition, less well-mixed emissions such as the various aerosols and indeed contrails. For an interesting discussion of the issues visit the 'realclimate' web site.[13]

Using this information, if the cumulative total of emissions released between 2000 and 2006 are subtracted from the midrange cumulative budget for the twenty-first century, this leaves the cumulative budget remaining for 2007 to 2100.

The next stage is to ask by how much emissions must be reduced to ensure the cumulative budget is not exceeded if emissions were to begin to decline at a constant rate in, say, 2008. The answer to this is a global emission reduction of some 55% by 2050; however, this figure is of little relevance when taking a more practical approach. Given the current trends in energy consumption, and with the considerable momentum to be reversed in making the transition to a low-carbon global economy, the reality is emissions will continue to rise for some years to come. For CO_2 emissions to peak and begin to decline within the coming decade will require an unprecedented and global effort to both reduce energy consumption and decarbonise global energy supplies. Figure 3.2 illustrates one consequence of delaying action. The additional emissions released between 2007 and the point at which the emission decline curve is crossed, must be added to the

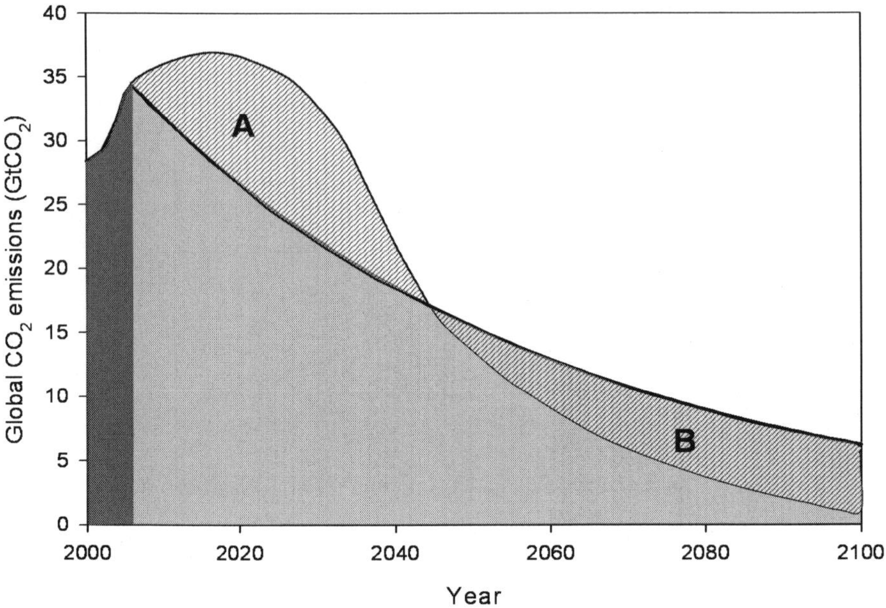

Figure 3.2 Illustration of the consequence in delaying emission reduction. The cumulative emissions A must be equal to B to remain within the same cumulative emission budget.

overall cumulative burden (A). To compensate for this, and ensure cumulative emissions are commensurate with the 450 ppmv level, emissions must reduce more rapidly than under the constant emission-reduction curve presented in Figure 3.1. The compensatory amount B must be equivalent to A to ensure the cumulative budget is not exceeded. In addition, the emission reduction in 2050 must now be greater than 55%.

The irrelevance of where those emission reductions take place can be argued—be it a reduction at a point source in a particular geographical location, or a sectoral reduction, such as from the global shipping sector. Nevertheless, it is clear that for a reasonable chance of not exceeding the 2°C level, significant reductions are required globally, with the likelihood being that all sectors and nations will be required to make their contribution. If the simple illustration of a constant emission reduction from 2007 is applied for a second time, but now in relation to OECD[14] versus non-OECD nations, the consequence of one group of nations reducing emissions by some 40% instead of 55% by 2050 is that the other must compensate and reduce by some 70%. One nation or sector may be required to compensate for another, to ensure emissions are reduced in an efficient way, and in the case of nations, to allow some countries to industrialise and improve their standards of living. However, if many sectors and nations wish to grow economically during the coming decades, a radical departure from the traditional link between GDP growth and energy consumption and/or carbon emissions will be required to make the reductions necessary.

From a sectoral perspective, arguments have already begun in relation to certain sectors being privileged over others. This is certainly true with respect to aviation, where a large proportion of aviation emissions have been omitted from, for example, the UK Government's carbon-reduction target.[15] Although a case for a much greater proportion of the emissions space for aviation may be made on the basis of inherent technical constraints to decarbonise aviation (see Chapter 4), such an argument will run into difficulties if, for example, the shipping industry requests the same treatment. The reality of the matter is that in terms of CO_2 emissions, there will unlikely be enough 'pie' to share out if a number of sectors request special consideration. These issues will be discussed in Chapter 7.

Cumulative emissions are not the only important factor in achieving a particular global stabilisation level of CO_2 emissions; the rate of increase in emissions may itself modify not only the ultimate CO_2 concentration attained, but also play a role in the associated temperature response. However, further research is currently underway to assess the scientific importance of rates of emissions growth. Nevertheless, for the purpose of informing urgent action on global climate policy, the crucial significance of the cumulative approach can not be overemphasised.

According to the IPCC, CO_2 is "the most important anthropogenic greenhouse gas".[16] Following the 'correlation trail' from global mean temperature rises and CO_2 concentrations through to global, then regional and national emission pathways, regions and nations can play a role in global CO_2 mitigation based on quantifiable carbon budgets.[17] In making the link between temperatures and carbon budgets, two key areas affected by scientific uncertainty are central to the debate, and have a considerable impact on any derived regional or national climate policy. One is *Climate Sensitivity*, and the other is the *Carbon-Cycle*.

Climate Sensitivity

Climate sensitivity is a measure of the climate system response to sustained radiative forcing.[16] It is usually defined as the global surface warming following a doubling of CO_2 concentrations from pre-industrial levels of 280 ppmv. The likely range, as published in 2007 by the IPCC, is between 2°C and 4.5°C with a best estimate of 3°C. In other words, if global concentrations of greenhouse gases stabilise at 560 ppmv CO_2 equivalent at some point in the future, then according to the IPCC, it is "likely" that global mean surface temperatures will ultimately rise by 3°C above preindustrial levels.

To put these temperature rises into context, the global and regional impacts of a 2°C rise include the destruction of the vast majority of coral reefs, three billion people experiencing water stress and changes in global cereal production that could expose as many as 220 million more people to the risk of hunger. At 3°C, few ecosystems would be able to adapt and, for example, there would be much larger losses in global cereal production than predicted at 2°C, potentially exposing a further 400 million people to hunger.[18] Clearly, the lower the temperature rise, the less significant the impacts likely to be suffered.

Given that the science relating to climate sensitivity currently provides guidance as to the global CO_2 concentration to be aimed for, the next consideration relates to the emissions budget associated with a related CO_2 concentration. This is where the carbon-cycle feedback studies play a clear role.

Carbon-Cycle Feedbacks

The atmospheric concentration of CO_2 depends not only on the quantity of CO_2 emitted into the atmosphere (natural and anthropogenic), but also on climate change–induced changes in the strength of carbon sinks within the ocean and biosphere. For example, as the atmospheric concentration of CO_2 increases (at least within reasonable bounds), so there is a net increase in the uptake of CO_2 from the atmosphere by vegetation (carbon fertilisation). Changes in temperature and rainfall induced by increased CO_2 affect the absorptive capacity of natural sinks and alter the geographical distribution of vegetation and hence its ability to store CO_2.[9] An increasing temperature speeds up the rate of decomposition of carbon and hence decreases storage capacity of the land.

The complicated and interactive nature of these effects leads to uncertainties with regard to the size of carbon-cycle feedbacks.[19] Nevertheless, the implications of carbon-cycle feedbacks for climate policy are profound. To achieve a desired CO_2 stabilisation level, the cumulative CO_2 emissions must be within certain bounds. Putting a value on these cumulative emissions is highly dependent on the degree to which carbon-cycle feedbacks are included within a particular model. Results of recent research show that the carbon budget available is reduced significantly when carbon-cycle feedbacks are included within models.[20] The consequences for national climate policies of including the results of carbon-cycle feedback studies are illustrated within this chapter.

Implications of Sensitivity and Carbon-Cycle Understanding

It is possible to link global temperature increases to particular stabilisation levels of atmospheric CO_2 concentration. However, in making this link, the uncertainty in relation to climate sensitivity must be taken into account. In the past, a doubling of CO_2 concentration from pre-industrial levels was considered to result in around a 2 to 2.5°C temperature rise. It is now considered that 450 ppmv is more closely aligned with this temperature rise than the higher 550 ppmv level.

Having established that it is necessary to aim for a much lower CO_2 concentration, if remaining within the 2°C threshold continues to be the ultimate goal, it is essential for regions and nations to understand their role in its achievement. This is where carbon budgets provide a firm scientific basis for developing national and regional carbon mitigation policy. Scientists can estimate the quantity of CO_2 emissions that can be released over a given time frame and relate this to stabilisation at the global atmospheric CO_2 concentration level. The uncertainty involved in this case is concerned with the carbon-cycle feedback effects. However, the IPCC 2007 report provides an improved understanding of those feedbacks relative to the past, the result of which reduces the global emission budgets that correspond with particular stabilisation levels due to the influence of positive carbon-cycle feedbacks. Consequently, regions and nations now face a much more challenging task than was previously thought. Firstly, in aiming for a 450 ppmv rather than 550 ppmv level, and secondly, being faced with significantly reduced carbon budgets due to the improved understanding of carbon-cycle feedbacks. The repercussions of this evolving global evidence-base for the EU and UK are addressed in the next sections.

CLIMATE TARGETS IN THE EU

As discussed in the previous section, a future global temperature threshold can be 'correlated' with a range of future atmospheric CO_2 concentrations and subsequently linked to a range of global cumulative emission budgets. The application of an apportionment regime to this range of global budgets delivers national or regional emission budgets for a given period. From the national or regional cumulative emission budget, national or regional emission pathways can be generated describing alternative future pathways lying within a nation's cumulative emission budget. Within this chapter, the 'correlation trail' from global temperature changes to national and regional carbon emission pathways is presented for the EU and the UK. This is done both for the older scientific understanding and the most recent global cumulative carbon budgets, with both analyses using the apportionment regime underpinning the UK Government's long-term carbon-reduction target.

The EU has adopted a target of global mean surface temperatures not exceeding a 2°C rise above preindustrial levels.[1] This target is now associated with stabilising atmospheric CO_2 at below 450 ppmv as discussed in the previous section. There are a number of important issues to be addressed in relation to the EU's climate change target and in turn how such targets relate to the aviation industry.

The starting point within the analysis is the ultimate aim of the target—i.e., for temperatures to not exceed the 2°C threshold. This threshold is associated with stabilising atmospheric CO_2 at a particular level, with each level having a different probability of exceeding 2°C. This approach demands all CO_2-producing sectors are included, as the atmosphere does not 'see' what is or is not accounted for. However, as is the case in the Kyoto Protocol, the EU target does not explicitly include the emissions generated by either international aviation or shipping. If these sectors contributed insignificant amounts of CO_2 both now and in the future, this may be a reasonable approximation. The analysis presented in this book and elsewhere illustrates, however, that this is not the case. To adequately devise climate policy for nations whose 'international' emissions are, or may in the future be, a significant proportion of their total, international sectors must be taken into account, and included within the budgeting. By taking a cumulative budget approach, it becomes clear that delaying action to mitigate emissions requires more stringent measures to avoid exceeding the 2°C threshold than is generally recognised.[21] One of the failures of climate change policy not based on cumulative emissions and neglecting important sectors is its excessive focus on the longer-term (e.g., 60% 2050 targets). Consequently, the debate tends towards issues of infrastructure related to energy supply (nuclear power, wind farms, etc.), when in fact it is the short- to medium-term, and hence energy demand, that is of crucial importance[22] in providing the necessary 'quick hits' to avoid emissions rising further.

Emission Apportionment

Even before Russia's signing of the Kyoto Protocol, debate had begun as to the form, scale and responsibilities of any post-Kyoto emission-reduction strategies. Whatever the form such strategies may take, significant reductions will not be possible without adequate commitment from all the high-emission nations, including the United States. The central complaint of the US Federal Government is that the Kyoto agreement exempts much of the world, including major population centres such as China and India, from compliance. Consequently, they see it as unfairly harmful to the US economy and therefore are unlikely to commit to any post-Kyoto agreement that does not require early participation by industrialising nations as well as industrialised.

One framework designed to respond to this inclusivity issue involves establishing permissible global CO_2 emissions and apportioning these to all nation states according to a particular and agreed set of rules designed to maintain temperatures within certain bounds. In doing so, all nations are required to design a carbon emission strategy tailored to their particular circumstances, in the knowledge that other nations are doing likewise. An emission-reduction regime requiring all nations to set targets from the start of the process, and that has gained some popularity over recent years, is the Global Commons Institute (GCI) Contraction & Convergence regime.[23] The GCI was founded in 1990 with a "Focus on the protection of the global commons of the global climate system". Since 1996, the GCI has encouraged awareness of the Contraction & Convergence regime as, so they contend, a practical interpretation of the philosophical principle that "every adult on the planet has an equal right to emit greenhouse gases". Contraction & Convergence is claimed to provide an international and equitable framework for arresting global anthropogenic emissions, with all nations working together to establish and achieve an overall yearly emissions target—contraction. Moreover, all nations converge towards equal per capita emissions by a specified year—convergence. By simultaneously contracting and converging, this mechanism requires all nations to impose targets from the outset, although for some nations this target may permit increases in emissions in the early years.

In addition to the Contraction & Convergence policy, with an a priori presumption that nations should move towards per capita equity in their CO_2 emissions, there are a number of related but different approaches to extending national commitments post-Kyoto. A study commissioned by the Umweltbundesamt (German Federal Environment Ministry)[24] assessed these approaches,[25] including Contraction & Convergence, and makes recommendations for increasing their effectiveness and acceptability. Some comments drawing on their findings on alternative approaches to Contraction & Convergence are summarised here:

- *Intensity targets:* currently favoured by the Chinese administration. These can play a role in future commitments as one form of target for a particular group of countries, possibly in parallel to other types of targets for other countries. If applied to all countries, the global emission intensity (emissions per unit of GDP) has to decrease rapidly in order to reach stringent environmental goals (particularly if economic growth is very high). Agreeing on differentiated intensity reductions may be more difficult than agreeing on the level of absolute emissions reductions, as emissions intensity involves country-specific knowledge of the relationship between emissions and GDP, which also may evolve with time.
- *Contraction & Convergence:* applied by the Royal Commission on Environmental Pollution (RCEP) in 2002 to calculate the UK's 60% carbon-reduction target since adopted by UK Government. As major reductions in emissions are necessary to avoiding 'dangerous climate change' it is likely per capita emissions under any policy regime will eventually converge to a very low level. Contraction & Convergence has the advantages of simplicity and stringency but does not account for the structural differences of countries or their ability to decrease their emissions.
- *The Triptych approach:* country-specific emissions budgets are calculated that reflect the energy, industrial and household sectors. As the method takes into account existing differences between countries, it can differentiate national emission-reduction targets based on need.

- *Multistage approaches:* "will be the future of the climate regime", but there are many possibilities regarding types of stages and thresholds for moving into a next stage. The current two stages (Annex I and Non Annex I) could be extended. One criterion for moving to a further stage could be emissions per capita.
- *The multisector convergence approach:* describes a complete set of rules for a future climate regime, defining in essence the path on which sectoral per capita emissions converge. A major downside of the approach is that sectoral activities are not necessarily directly related to optimal population levels.
- *Equal mitigation costs:* setting targets so that mitigation costs are equal for all participating countries (e.g., a percentage share of the GDP), has the theoretical potential to be a 'fair' option. In practice, however, it may be impossible to agree on a model or calculation method for calculating the cost of countries.

Other policy approaches could be added to the list, notably the Brazilian approach in which emissions reduction responsibilities are allocated on the basis of countries' historical contribution to global temperature change.

Contraction & Convergence has the advantage of simplicity, an element of international equity, includes all nations thereby encouraging early action from the outset and avoids carbon leakage (where emissions increase in areas or nations not covered by a regime). However, it does not allow for structural differences between countries that affect their ability to reduce emissions. Furthermore, given the scale of the challenge faced, only a few of the least industrialised nations will likely be able to trade emission allowances with industrialised nations in the short- to medium-term.[26] This is, however, likely to be similar for the other regimes available given the evolving scientific evidence-base relating to climate sensitivity and carbon-cycle feedbacks already described.

The EU Scale: Carbon Budget Analysis

Despite Contraction & Convergence being one of several options for a post-Kyoto climate regime, given its importance and popularity within the UK and EU, the analysis of regional and national carbon budgets within this book is carried out using this particular policy regime. However, as will be explained, this approach will be used in two ways, firstly in its standard form making use of the GCI's Contraction & Convergence model, and secondly using the aggregated emission budgets produced using the Contraction & Convergence approach but deriving consequent emission pathways separately. The analysis is structured in this way to further illustrate the importance of ensuring policy measures keep up-to-date with the evolving scientific debate; the first approach is more closely aligned to existing EU and UK Government climate policy, whilst the second is more scientifically credible.

As it currently stands, the Contraction & Convergence model developed by the GCI provides policy guidance in relation to CO_2 targets for the period 2000–2100. By choosing a convergence year—the year in which nations reach equal per capita emissions, and a twenty-first-century carbon budget in line with a desirable CO_2 concentration, the model will output national emission pathways from the year 2000 to 2100. When this analysis was first carried out,[27] the cumulative budgets used in the model were based on earlier IPCC results in which carbon-cycle feedbacks were not included. This is also the model version used when the RCEP derived the UK's 60% carbon-reduction target. The resulting EU emission pathways for both the 550 ppmv and 450 ppmv CO_2 levels are presented in Figure 3.3.

Under this regime, for the EU to play its fair role in achieving a 550 ppmv CO_2 stabilisation, the whole of the EU needs to reduce emissions from around 1120MtC in 2000 to close to 450MtC by 2050—about a 60% reduction. If the target is commensurate with 450 ppmv CO_2, the equivalent reduction is to 217MtC; an 80% reduction.

However, there are a number of problems with the approach described above. Firstly, the IPCC has published new cumulative CO_2 emission data for the twenty-first century incorporating the latest understanding of carbon-cycle feedbacks.[28] Secondly, the 550 ppmv CO_2 stabilisation level provides a very high risk of exceeding 2°C (Meinhausen 2006); therefore

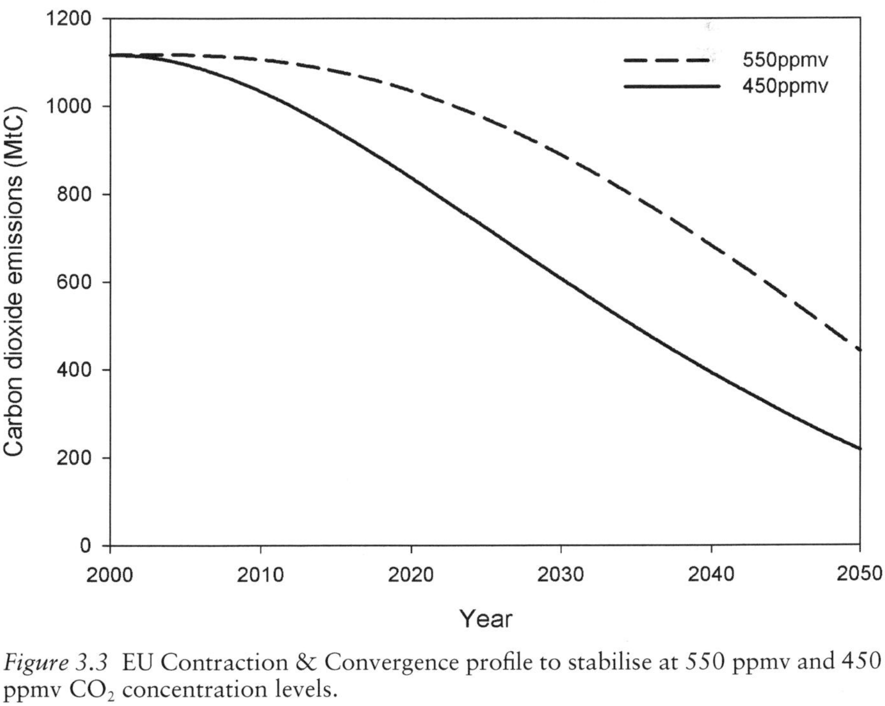

Figure 3.3 EU Contraction & Convergence profile to stabilise at 550 ppmv and 450 ppmv CO_2 concentration levels.

further analysis will concentrate on the 450 ppmv level only. Thirdly, the GCI's model does not include 'bunker' emissions. Bunker emissions are those CO_2 emissions released by international aviation and shipping. These are considered to be some 3–5% of EU CO_2 emissions, although currently there is much debate as to the correct shipping value. Finally, the emission profiles produced by the model do not take into account emission increases from 2000–2007. As cumulative emission budgets are central to stabilising CO_2 concentrations, discrepancies between modelled emission reductions from the year 2000 compared with actual emission growth must be taken into account.

An alternative approach to attempt to rectify these problems and provide a more innovative and appropriate analysis is therefore highly desirable. To address the points highlighted in the previous paragraph, the following approach is employed and data used:

1. The analysis is based on the IPCC's 2007 range of cumulative emission values for the twenty-first century.
2. There is a focus on stabilisation at the 450 ppmv CO_2 level.
3. Global bunker fuel emissions are included.
4. Emission pathways are derived from an understanding of current emissions, trends and drivers; the GCI approach only provides the regional emission budget.

The results of conducting the analysis on the basis outlined above are demonstrated graphically in Figure 3.4 with the cumulative values stated in Table 3.1.

Table 3.1 summarises the cumulative carbon budgets at both the global and apportioned EU scales for the high and low range published in IPCC 2007. To produce emission trajectories for the EU25, those emissions already recorded for 2000–2004 make up the first part of the pathway (Figure 3.4). At the time of the analysis, 2004 was the latest year for which data was available. Despite being just four years of data, the cumulative emissions over this period are a significant portion of the overall emission budget over a 2000–2050 period. For example, in the case of '450 High', these emissions represent 14% of the total fifty-year budget. In other words, EU nations are spending their carbon budgets very rapidly. Given the budget available, the remainder of the pathway is developed by choosing a realistic but challenging date for emissions to peak, and subsequently creating pathways that remain within the constrained budget. Clearly, the higher the cumulative target, the easier it is to manoeuvre in later years, with the converse being true for lower cumulative targets. Hence any policy aiming to stabilise CO_2 concentrations in line with either '450 Low' or '450 High' must stabilise emissions as a matter of some urgency, and maintain significant year-on-year reductions for three decades. A comparison of the EU25's emission pathway with possible aviation futures will form the basis for Chapter 6.

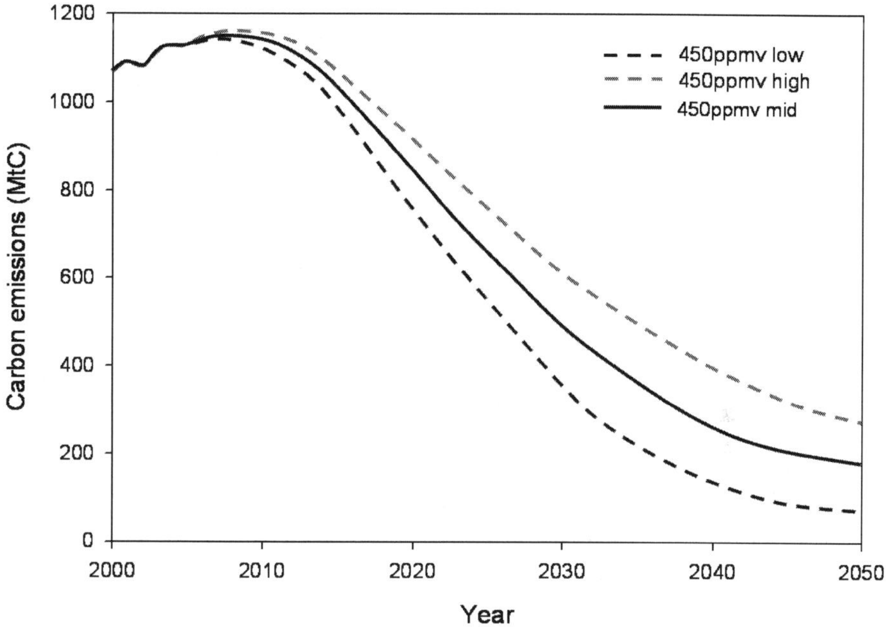

Figure 3.4 EU25 CO_2 emission pathways consistent with the cumulative carbon budget published by the IPCC in 2007. Emission for the years 2000–2004 are taken from submissions to the UNFCCC.

The UK Scale: Carbon Budget Analysis

In February 2003, the UK's Department of Trade and Industry (DTI) published *Our Energy Future—Creating a Low Carbon Economy*.[4] The White Paper essentially accepted the analysis of the RCEP in their twenty-second report *Energy—The Changing Climate*. The RCEP argued that a "Contraction & Convergence" policy was required for international control of carbon emissions, a consequence of which is a requirement for a 60–90% reduction in carbon emissions by industrialised countries. The principal objective of the RCEP (and, by association, the Energy White Paper) is to avoid "dangerous climate change" by ensuring the global mean atmospheric concentration of CO_2 does not exceed a particular stabilisation level. This

Table 3.1 Global and EU Cumulative Carbon Budgets for 1990–2100

Scenario	*Global Cumulative Emissions (GtC) (2000–2100)*	*EU Cumulative Emission (GtC) (2000–2050)*
450 Low	375	30
450 High	600	39

was understood by the Government and RCEP as being consistent with the goal of the Framework Convention on Climate Change.[29] The 2003 Energy White Paper thus set a target of reducing total UK CO_2 emissions by 60% from 'current' levels by 2050 as shown in Figure 3.5, in line with stabilisation at 550 ppmv, the level now thought to have a very high probability of exceeding the 2°C threshold.

Taking the identical approach to that described in relation to the EU25, a more refined analysis of the UK's carbon budget can be made in relation to a 450 ppmv CO_2 stabilisation level, and incorporating emissions from international aviation and shipping. In this case, because international aviation emissions are a greater proportion of the UK's total than they are for the EU, as outlined in Chapter 2, the emission pathway is more challenging than the one for the EU as a whole. The new emission pathway range for the UK is presented in Figure 3.6. In this case, the scenarios are identical between 2006 and 2012, in line with an assumption that all non-international sectors could stabilise their CO_2 emissions, if so encouraged. CO_2 emissions from international aviation and shipping, being the two fastest growing sectors in terms of CO_2, are assumed to continue to grow, but at rates marginally lower than current trends.

The emission reductions required of the UK to remain within budget are extremely challenging, as illustrated in Figure 3.6, with annual reductions of

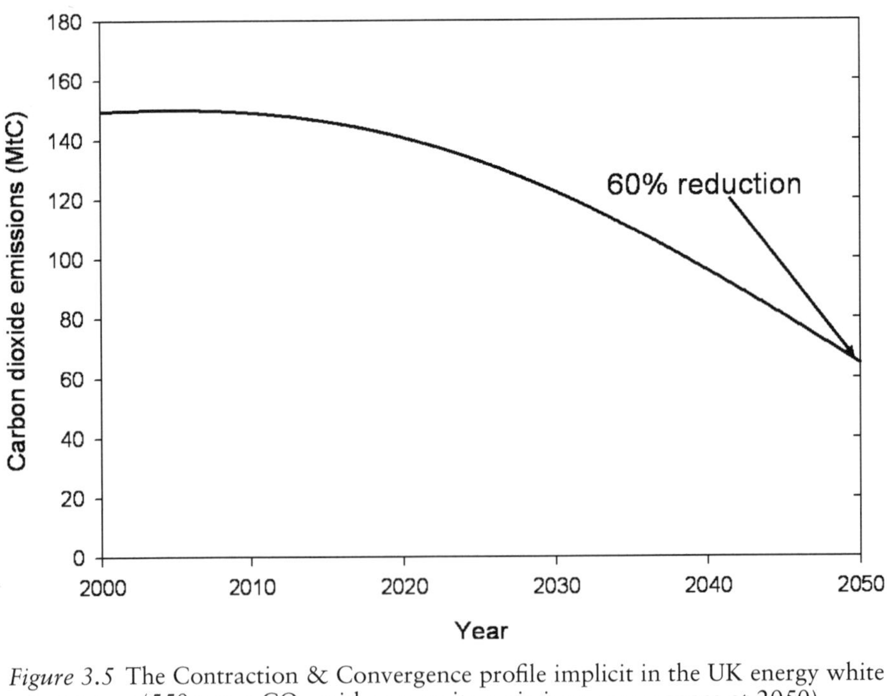

Figure 3.5 The Contraction & Convergence profile implicit in the UK energy white paper target (550 ppmv CO_2, with per capita emissions convergence at 2050).

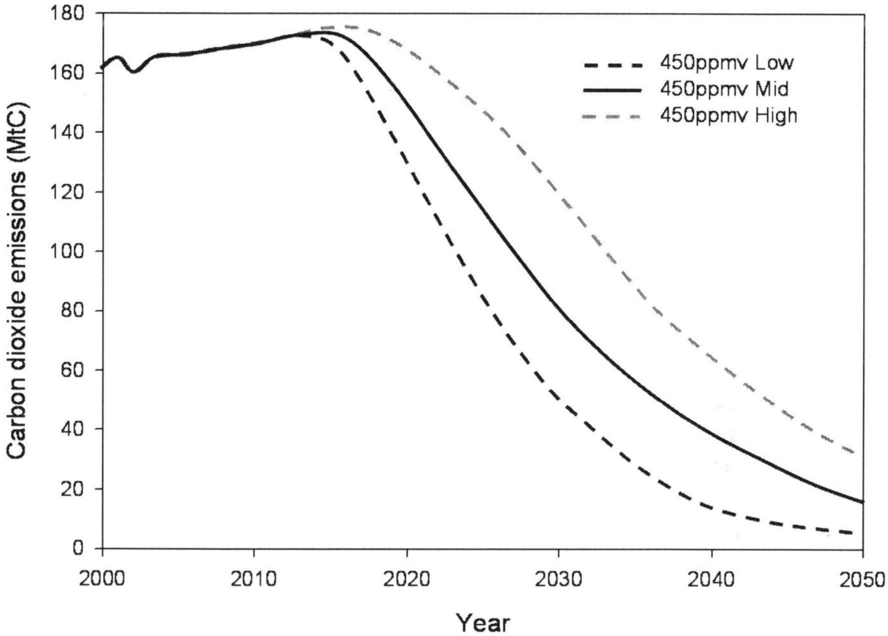

Figure 3.6 CO_2 emission pathways for the UK for a 450 ppmv stabilisation level.

around 6% between 2020 and 2030 for the midrange scenario. These emission reductions are much greater than UK policymakers are currently considering. Furthermore, when incorporating economic growth rates, annual carbon intensity reductions are of the order of 8% per year, assuming a 2% per year growth in the UK economy. Only when governments engage with the very deep annual cuts required of industrialised nations to remain within the 2°C threshold, will effective measures and regulations be developed that are able to successfully tackle the issue at hand.

SUMMARY

The beginning of this chapter presents some of the key issues in relation to climate change. It goes on to stress the importance of regional and national climate change policy keeping up-to-date with the evolving scientific understanding of climate change. Two areas of recent development crucial to the formation of climate change policy are a refined understanding of climate sensitivity and carbon-cycle feedbacks. Recent modifications to both result in a lower global carbon budget associated with the 2°C threshold. Implications for regional and national climate policy are profound, and only serve to increase significantly the importance of urgent action to mitigate emissions.

In the second section of the chapter, the new global CO_2 budget range is used to quantify regional and national CO_2 emission pathways between 2000 and 2050. Prior to the publication of the *Stern Review*[30] of the economics of climate change, and Tyndall's *Living Within a Carbon Budget*,[31] UK energy and climate change policy documents generally neglected the importance of cumulative emissions, and their impact upon CO_2 stabilisation levels. Whilst Stern considers a global cumulative emission trajectory, the analysis presented here is the first to consider cumulative emissions in terms of the UK's and EU's carbon budgets, based on an understanding of current energy policy. Results point to carbon emission pathways very much more challenging than current policymakers within the UK and EU are willing to openly accept. Unless such policymakers are prepared to engage with the true scale of the challenge when developing policies, the 2°C target will remain little more than an uninformed desire of well-meaning politicians. Such a situation cannot be reconciled with evidence-based carbon mitigation and adaptation policy.

4 Opportunities for Aviation

INTRODUCTION

The previous chapter presented the scale of the challenge faced by all nations and all sectors in mitigating CO_2 emissions and addressing climate change. The aviation sector cannot be an exception given the scale and rate of growth of its emissions coupled with the urgency and magnitude of the action required to avoid 'dangerous climate change'. This chapter discusses the current and future technological and operational opportunities within the aviation sector to improve fuel efficiency and carbon intensity. Although it is widely accepted there will be no step change in technology to bring about a reduction in emissions in real terms within the coming decade, nevertheless, new momentum to recognise and address the urgent climate change issue may now be accelerating innovation within the aviation industry. One indisputable issue of specific relevance to this sector is the long lag in bringing about change given the lifetime of current aircraft. Furthermore, those aircraft designs currently coming into service began their developmental life before environmental pressures rose significantly up the political agenda.[1] This chapter summarises some of the technological and operational options available for improving the efficiency and carbon intensity of aviation, potential deployment timescales and the likelihood or otherwise of their early adoption.

SHIFTING ENVIRONMENTAL FOCUS

Minimising the noise and local air quality impacts of the aviation industry has been of significant importance to the industry for many years, not least in response to grassroots opposition at a number of large hub airports such as Heathrow. In addition, for economic reasons, improving fuel efficiency has increasingly become important to the aviation industry, with continued improvements to aircraft engines, airframe designs and operations leading to annual reductions in the fuel consumed per passenger-km. The principal driver to reduce the amount of kerosene consumed has traditionally been business efficiency—fuel is a significant cost to airlines, therefore reducing the outlay to fuel likely improves profitability. More recently however, climate change has risen up the political agenda, particularly in the UK and EU, therefore additional drivers to reduce fuel consumption, and hence CO_2 emissions, are coming to the fore. Unfortunately, striving to reduce the environmental impact of the aviation industry is not straightforward. Aside from a reduction in flights and passenger-km, win-win solutions to noise, local pollution and CO_2 and other climate-warming emissions are relatively scarce. For example, noise-reducing equipment on aircraft increases their weight, and hence the fuel consumed, particularly during takeoff and landing. Similarly, increasing the operating temperature and hence fuel efficiency of jet engines generally results in increased NO_x emissions and hence increases the impacts on local air quality. Nevertheless, the industry is continually seeking new methods and techniques for reducing their environmental impact.

Just five years ago, if the term 'environment' was used in relation to aviation, it would generally be assumed to relate to noise or local pollution. By 2007, the environmental focus on aviation, particularly in the UK and the EU, has shifted towards climate change. Pressure for the aviation industry to address its own climate change impact came from a variety of sources,[2] but was particularly prominent in a number of academic papers and reports from the Tyndall Centre.[3] At a similar time, and responding to additional pressure from the EU Commission, ACARE published targets encouraging the aviation industry to play its part in line with the Kyoto agreement.

More recently, to improve on the measures outlined by ACARE, the EU is driving forward proposals to include the aviation sector within the EU's Emissions Trading Scheme. In an era of enhanced airport security coupled with a regular stream of climate change messages coming through the various print and broadcast media, consumer pressure to either substitute mode of travel, or choose the 'greenest' airline, is arguably gaining strength. In combination, the current state of affairs appears to be accelerating efforts internal and external to the industry to minimise aviation's climate change impacts.

The ACARE targets are asking for an 80% reduction in NO_x emissions for a new plane in 2020 compared with an equivalent new plane in 2000, plus a reduction of 50% per seat-km in CO_2. These targets are considered within the industry to be challenging, but not unattainable. To achieve this 50% reduction in CO_2 and hence fuel burn per passenger-km, 10% is expected to be delivered through improvements to operations, 20% through improved propulsion efficiency and 20% through aircraft manufacture. To explore how these changes might be brought about, the remainder of the chapter will describe the various technical and operational solutions either currently available or on the horizon for reducing CO_2 and NO_x emissions and for minimising contrail and cirrus cloud production. In addition, opportunities for widespread use of low-carbon fuels will be explored.

AIRCRAFT ENGINE TECHNOLOGY

Aircraft engine manufacturers have continually worked to reduce the fuel consumption per passenger-km from aircraft engines, and have made significant improvements over the decades. To address the NO_x ACARE targets, areas of research funded by the industry currently include projects on lean burn combustor technology and investigations of how to improve heat management, combustion, active systems and core components of aero-engines. Further improvements to engine pressure ratios and reductions in NO_x emissions during the cruise stage of flight may be available, although could lead to additional fuel burn. However, confidence in achieving the ACARE NO_x targets through new and improved technology is high.

The most fuel-efficient engines for today's aircraft are high-bypass, high-pressure ratio gas turbine engines. These engines have high combustion pressures and temperatures, which are features consistent with good fuel efficiency, but generally this is at the expense of increased NO_x emissions—especially at high-power takeoff and at altitude cruise conditions. Although jet technology is relatively mature, reducing fuel burn in engines will continue to improve through advanced combustion research. For example, the Rolls-Royce Trent 1000 currently being developed for 2008 may be 12% more fuel efficient than a similar engine entering service in the year 2000. Further gains can be made through the use of open-rotor engines as opposed to high-bypass ratio turbofans. This type of engine would likely be used for short- to medium-haul aircraft primarily. Overall, engine technology is likely to continue to improve incrementally, leading to moderate improvements in fuel efficiency, together with significant reductions in NO_x emissions over the next fifteen years.

AIRFRAME DESIGN

The design and physical shape of aircraft can have a big impact on the level of drag experienced and hence fuel burn, as aircraft travel through the air at high speeds. Novel aircraft designs for reducing drag have been investigated: for example, the blended wing-body (BWB) aircraft and the wing-in-ground effect vehicle (WIG). In contrast, the latest aircraft being built by Boeing and Airbus, the two largest aircraft manufacturers, continue to use conventional airframe designs. Indeed the RCEP's aviation report[4] states that aircraft designs up to 2030 are thought likely to be based around conventional airframe configurations, but integrating best practice technology. However, the *Greener by Design Annual Report 2006–2007*[5] suggests that if the basic configuration is retained, then the prospects for further significant drag reduction are small. Given manufacturers are urged by ACARE to deliver a 20% fuel burn improvement by 2020, an acceleration in research and development driven by a carbon-constrained world may lead to new designs that currently exist only in the prototype or concept stages. Some of the most likely developments are highlighted in the following.

Blended Wing-Body Aircraft

The BWB or 'Flying Wing' design has a long history, with precedents in the German Horten aircraft AW-52 and the Northrop YB-49. The BWB has the body partly or wholly contained within the wing so that the interior of the wing in the central part of the aircraft becomes a wide passenger cabin. It could, its proponents claim, be significantly lighter and experience very much lower drag than the conventional swept wing–fuselage airframe design. Its fuel usage would therefore be reduced, perhaps by as much as 30%, further reducing aircraft takeoff weight. This type of aircraft would have a lower cruise altitude and an extended optimal range, because of the lower weight and drag. Some studies suggest a large commercial BWB aircraft could be developed to carry 800 or more passengers. It is thought that a BWB airliner cruising at high subsonic speeds on flights of up to 7,000 nautical miles would have a wingspan slightly wider than a Boeing 747 and could thus operate from existing airport terminals.

It is likely that the BWB concept will be applicable only to relatively large aircraft, as the embedded passenger cabin must be tall enough to enable passengers to stand up, so implying the need for large wings. Currently, it is not considered likely that BWB aircraft could represent a significant proportion of aircraft movements for a number of decades.[4]

Boundary Layer Laminar Flow

Improving the laminar flow (the flow of air over the aircraft) can significantly reduce drag, and hence improve fuel burn. Hybrid laminar flow control has been recently demonstrated on an Airbus A320 by applying suction to the fin. Natural laminar flow control, though simpler and operating without suction, is applicable only to small to medium sized aircraft and limited to wings with low sweep and therefore lower speed. By contrast, the hybrid laminar flow control involving boundary layer suction over various parts of the aerodynamic surfaces is more applicable to large aircraft.[5]

Airships

A very different approach to the problem of reducing the climate impact of aviation is to look at entirely different methods of air transportation. One such form is the airship. Modern airship designs use helium as a much safer alternative to hydrogen, which was used historically in the zeppelin fleet. Helium is heavier and more expensive to produce than hydrogen, and consequently additional lifting power may be required on takeoff (10% of lift is lost relative to a hydrogen filled airship).

A review of the potential for airships[6] concluded that tasks such as surveillance, airborne early warning (replacing Airborne Warning and Control System aircraft (AWACS)) and long tourist trips are better suited to airships than aeroplanes and helicopters. Small airships, such as the Zeppelin NT, are currently in operation in this capacity, although do not operate economically. On the other hand, larger volume craft are more likely to be profitable.[7] In relation to the feasibility of airship freighters, despite being significantly less damaging to the climate than a conventional jet aircraft, one study concluded that their use was 'unpromising' due primarily to manoeuvrability difficulties in wind during the loading and unloading stages. Offloading can occur in two ways:

1. The airship hovers where lateral movement (known as drift variation) is less than 1–2% of the vehicle length. Achieving this low level of drift variation is very difficult with such a large surface area against which wind and thermal forces act.
2. The airship descends and is moored to a specially built platform on land or water, although there is still the danger of capsizing in strong side winds.

The German Cargolifter airship was designed with the intention of hauling up to 160 tonnes for distances of as far as 10,000 km, but the project failed due to difficulties with financial and engineering management.

One of the most promising recent designs for a cargo lifter was the Skycat by Airship Technologies Group UK[6] but this company too become insolvent in July 2005—another setback for the future of airships and illustrating the economic difficulties of introducing step change to the technology in current markets. To date, no successful large cargo lifter has been built, even though major firms (such as Lockheed) have planned projects.

Wing-In-Ground Effect Vehicles (WIGs)

Aerodynamic drag on aircraft can be divided into two categories—that caused by the vortices around the wings (induced drag) and that due to the surface friction. As the distance between the ground and the wing decreases to a length less than an aircraft's wingspan, the ratio of lift to drag increases—this is known as 'ground effect'. For smaller aircraft, the increase in surface friction drag due to the denser air at lower altitudes is of roughly equal magnitude to the decrease in induced drag and so any fuel benefit is lost. For large vehicles however, such as the Boeing Pelican concept,[8] a much larger payload can be transported for a given range than for flight at conventional altitudes, or a given payload can be transported further with equivalent fuel.

The Pelican aircraft would have a wingspan of 150 m, would fly as low as 6 m above sea level and carry a load of 750 tonnes of cargo for 18,500 km when in 'ground effect' above the sea. At more standard altitude levels, this range for the same fuel burn would be reduced to 12,000 km. Operating such a large, heavy aircraft from existing airport facilities is unlikely. Furthermore, its maximum speed would be lower at low altitude due to air density; therefore the aircraft would take longer to reach their destinations than do current cargo aircraft. There may also be a problem with the certification of transoceanic flight at low altitude, as this would not fit into any current regulatory framework. From the noise point of view, the Pelican has a significant disadvantage over conventional aircraft. Its proposed high number of separate undercarriages would create much more noise on takeoff and landing than its conventional equivalent. If the WIG design is to prove successful, it is likely it will do so first as a freighter rather than a passenger aircraft.

Materials

Improvements to the materials used to construct aircraft can bring substantial benefits in terms of reducing aircraft weight and hence fuel burn. The majority of aircraft in the world fleet are built primarily from aluminium. However, by 2020, it is likely that many aircraft will be made up of large proportions of carbon composite materials. Indeed the Boeing 787, due to begin service in 2008, and the Airbus A350 will have about 50% carbon composite structure, offering major improvements in terms of fuel efficiency. As these aircraft are purchased and replace older more inefficient aircraft, as long as those older aircraft are retired, substantial improvements to fuel efficiency per passenger-km are likely. It is hoped that carbon composite aircraft will offer around a 10% contribution to the 50% ACARE target by 2020.

LOW-CARBON FUELS

Most other sectors of the economy can make use of, or are already using low-carbon fuel sources. The UK's electricity grid, for example, incorporates growing amounts of renewable energy, hybrid petrol-electric cars are on the market and some homes and businesses consume lower-carbon electricity and heating from combined heat and power systems (CHP). Aviation, on the other hand, has been somewhat slower to respond to the fuel-shift challenge, choosing to focus much more on fuel efficiency, engine technology and operations. This is slowly changing, perhaps at an accelerated pace in very recent years, with biofuels in particular mooted as a possible low-carbon option for aviation.

Biodiesel

Biodiesel as an aviation fuel would likely take the form of a kerosene extender. That is, biodiesel would be mixed with mineral kerosene to produce a new, lower-carbon-emitting fuel. A maximum of 10–20% of biodiesel could be used in aviation fuel, as in higher proportions, biodiesel alters the crystallisation properties of the aviation fuel at low temperatures. Current research efforts focus on, for example, filtering techniques to remove such crystals in mixtures containing up to 10% biodiesel so that the fuel continues to meet safety requirements. However, further research will be required for mixtures containing more than 10% biodiesel.

Advantages of biodiesel over conventional kerosene include its lower polluting emissions, its biodegradable nature and its relatively simple production from major bio-crop feedstocks (e.g., rape seed, soya bean etc). One key problem, as implied above, is that mixing mineral kerosene with biodiesel compromises kerosene's ability to perform at low temperatures, such as those experienced at altitude, even when mixed with a small proportion of biodiesel.[9] Further research will be needed to improve and build confidence in cold weather performance. Moreover, conventional biofuel production on a large scale has considerable negative sustainability implications. Second and third generation biofuels may help to provide a more practical solution.

Bio-kerosene

As an alternative to biodiesel, kerosene can be manufactured from biomass via the Fischer-Tropsch chemical conversion process. This has the advantage of providing fuel-cycle CO_2 benefits compared with mineral kerosene, and eliminating oxides of sulphur. Bio-kerosene is chemically and physically similar to mineral kerosene, and therefore broadly compatible with current fuel storage and engines.[9] However, its lack of aromatic molecules and the fact that it is virtually sulphur-free, gives it poor lubricity. It also has a lower energy density than mineral kerosene, which would impact on long-haul flights. A few modifications could, on the other hand, improve its lubricity, making it fit for use. This type of kerosene is likely to be a medium-term development for the aviation industry, although again, sustainability concerns remain.

Hydrogen

The benefits of using hydrogen to fuel aircraft would be maximised if derived from the gasification of biomass or electrolysis of water using low-carbon based electricity. Ponater et al.[10] suggest that hydrogen has the potential to reduce aircraft-induced climate impact significantly. However, the first difficulty with using hydrogen as an energy carrier for aircraft is that it would require fundamental changes to the jet design. The high energy content, but low density of hydrogen requires much larger fuel tanks. Although there would be a weight advantage due to aircraft carrying lighter fuel, this would be offset to some degree by the weight of a larger fuel tank.

The volume of hydrogen carried would also be some 2.5 times that of the equivalent kerosene. The airframe would therefore need to be larger, and so would have a correspondingly larger drag. The combination of larger drag and lower weight would require flight at higher altitudes. Therefore, if and when hydrogen does come into use as an aviation fuel, it will most likely be used in large long-haul, high-altitude aircraft.[4] The requirement to carry a greater fuel volume may present an added difficulty for a hydrogen-fuelled blended wing-body aircraft, a design otherwise well suited to long-haul flights.

The effects of NO_x would still be present when using hydrogen as an aviation fuel, depending on the burn temperature. Perhaps of more concern would be the enhanced production of water vapour, which depending on the cruise altitude, could increase contrail formation and hence cirrus cloud. Aside from problems of hydrogen storage, transportation and the need for new infrastructure worldwide,[11] hydrogen's main by-product is water vapour—which acts as a greenhouse gas in the upper troposphere. Therefore, the sensitivity to cruising altitude is likely to be very large.[12] If, as appears likely, hydrogen-fuelled aircraft were to cruise at higher levels, then the increased water emitted into the stratosphere would likely produce larger radiative forcing. Since a hydrogen-fuelled aircraft would produce more water than a kerosene-fuelled aircraft, and since the water vapour produced by the latter cruising at 17–20 km gives a radiative forcing some five times that of a lower flying subsonic aircraft, a hydrogen-fuelled supersonic aircraft flying at stratospheric levels would be expected to have a radiative forcing some thirteen times larger than for a standard kerosene-fuelled subsonic aircraft.[4]

Further research would therefore be required to ensure that any advantage gained in reducing carbon emissions, would not be exacerbated by an increase in climate impacts due to enhanced water vapour production. Overall, the environmental benefits of using hydrogen rather than kerosene for fuelling aircraft engines are doubtful; the RCEP consider hydrogen as likely to be discounted as an aviation fuel for many decades.

Other Fuels

Other fuels considered and subsequently rejected to date include ethanol and methanol. Their very low heat content, in mass and volume terms, render them useless as jet propulsion fuels. Moreover, from a safety standpoint, these alcohols have very low flash points—12 and 18°C compared with the minimum standard allowed of 38°C.[9,11] Nuclear-powered aircraft are also not currently being considered due to safety concerns over radiation leaks and the implications of crashes. Finally, bio-methane has been considered as an alternative to kerosene, but it would require similar infrastructure and aircraft design changes to hydrogen.

Fuels Summary

To summarise, biodiesel and bio-kerosene could be used in conventional airframe designs and engines, but further research is required to make biodiesel of practical use in cold conditions, and bio-kerosene has large land-use and hence sustainability implications if conventional biofuel sources are to be exploited. Overall, bio-kerosene appears to be the most viable option in the medium term. Hydrogen, on the other hand, has questionable environmental advantages in this context and would require large-scale changes in terms of the supply infrastructure and airframe design. It is unlikely that hydrogen will be used to fuel planes for the foreseeable future.

Although it is often considered that many of the technically feasible low-carbon fuel options would likely be used in surface transport in preference to aviation, due to cost and ease of implementation, it may be environmentally advantageous to prioritise a limited supply of biofuel for the aviation sector. Surface transport modes have a number of options more readily open to them, including hydrogen fuel cells and electricity; whereas it may be the case that bio-kerosene is the only alternative fuel open to the aviation industry within the timescale required to address climate change. This issue merits additional research.

One final comment on the use of alternative fuels within aviation relates to the fact that only a few years ago (2004 to 2005), aviation industry representatives interviewed by the Tyndall Centre indicated they would be surprised to see alternative low-carbon fuels making significant inroads into the sector prior to 2030. This attitude has altered in only a short space of time, with biofuels now very much on the agenda. For example, Virgin Atlantic have recently joined forces with Boeing to test their first biofuel-powered aircraft in 2008, although it is currently unclear what type of biofuel and where the source of such fuel might come from. Significant biofuel usage may continue to be some years away; however this modification to the aviation industry viewpoint can only serve to encourage quicker take-up and roll-out of new technology.

OPERATIONS

Technical options for reducing the fuel consumed per passenger-km or freight-tonne-km are in general, incremental and slow to come to the fore. Operational options, on the other hand, could bring about reductions in fuel burn on a much shorter timescale. The drive for improved fuel efficiency therefore also needs to encompass airport and airline managerial aspects of the flight system. Evidence for how alternative operational practices offer gains in efficiency is clear from the success of the low-cost air models. Due to a combination of more efficient operational practices and appropriate acquisition and use of aircraft, Easyjet can currently legitimately claim to emit less CO_2 per passenger-km on average than legacy airlines such as British Airways. Principal components of the efficiency gains through operations come from higher load factors, higher seat density and fast turnaround times that reduce the time aircraft are burning fuel whilst stationary.

Load Factors and Seat Density

The load factor of an aircraft relates to how many seats on the aircraft are occupied. This should not be confused with seat density, which is a measure of how many seats are available within the aircraft. For example, a low-cost airline may purchase an identical aircraft to one owned by a legacy airline, but request a higher seat density from the manufacturer. Despite a slight weight gain for the aircraft, increasing the aircraft's capacity in this way leads to lower fuel costs and CO_2 and other emissions per passenger-km. Due to more differentiation within the seating available within the legacy airline business model, aircraft seat density will not be maximised. This is particularly true of medium- to long-haul aircraft. Coupled with this, a high load factor for those densely seated aircraft will further reduce the quantity of fuel spent per passenger, and reduce the need for as many flights, assuming passenger demand remains constant.

As there is a direct relationship between load factors and fuel costs, airlines continually strive to increase the numbers of passengers onboard. However, scheduled or legacy airlines will unlikely achieve the kinds of high percentages (~85%) of the likes of Easyjet due to differences between their business models and hence seat-pricing structures. Low-cost airline models not only sell seats for a lower price than legacy airlines, but they make a significant proportion of their profits from the sale of ancillary items, such as hotel bookings, car hire and theatre tickets, for example. Therefore, the relationship between the number of passengers and profits to be made is extremely strong. Only by very carefully managing the price of each seat sold (with some at very low and some at much higher prices) to both minimise the perceived seat cost to the passenger and maximise profit, can the airline be a success. Legacy airlines, on the other hand, have much less price differentiation between seats of similar class. The legacy business model does not necessarily wish to be seen to sell off tickets at a very low price as time goes on simply to fill the aircraft. Furthermore, given much on the aircraft will be complimentary, drinks, food and entertainment, etc., selling a seat for £1 with a chance of making no further money from the passenger, would probably not cover the cost of incurred additional administration.

Increases in seat density and load factors are possible across the global fleet, particularly if the legacy airlines adopt something more closely aligned to the low-cost model in this regard. In addition, more sophisticated ticketing systems, inter-airline alliances to merge flights at the last minute, differing pricing bands and demand-focussed time-tabling could provide further gains.

Air Traffic Management

Aircraft burn a substantial proportion of their fuel during takeoff and landing, which is why an indirect flight from Manchester to London then London to Madrid, has a much larger environmental footprint than a direct flight between Manchester and Madrid, for the same aircraft type. An increase in point-to-point flying rather than the commonly used hub-to-hub flights might reduce fuel consumption for a given route, though the additional weight of fuel carried on a direct route would need to be considered. Nevertheless, hubs are arguably principally used for economic reasons: they enable cross-subsidy of 'thin' routes that would otherwise be unprofitable. On the other hand, the new Airbus A380 superjumbo was designed to improve fuel efficiency at large hub airports by reducing the number of flights required for a given number of passengers. The environmental benefit or otherwise of the hub-and-spoke model versus point-to-point is not clear, and probably depends on the flight length and popularity of the route in question.

Capacity constraints are an additional issue that some hub airports are addressing partly through the purchase of the superjumbo A380. On the one hand, these capacity constraints may be stifling growth, which, from an environmental perspective, is positive. However, such constraints lead to increases in circling and taxiing time on runways, adding to the emissions burden. The A380 aircraft could benefit such hub airports in this respect, if the actual number of flights does not increase. However, if growth is then boosted as a consequence, the environmental gains may be negligible, or even negative. Furthermore, a greater distance is required for the wake of an A380 during takeoff and landing compared with smaller aircraft, which again has capacity implications. At the time of writing, the A380 had just been delivered to Singapore to begin commercial flights. Any rebound effects in terms of fuel gain or loss (per passenger-km) through the purchase and use of the A380 will become evident over the coming years.

Flight routes taken by aircraft can make a difference to the fuel consumed. Until very recently, aircraft have had to fly fixed routes that are an historic part of the infrastructure, resulting from the days when following a set of ground beacons was the only reliable source of navigation for aircraft. However, with the advent of global positioning satellites (GPS) and modern flight management systems on board airliners, it is now possible to derive a set of waypoints that are not necessarily linked to physical locations on the ground. These technologies enable the introduction of new concepts of operation, such as 'direct routing' whereby the aircraft determines an optimal flight path from the start to the destination airports without reference to fixed points on the ground.[13] Such improvements will translate directly into reductions in fuel consumption per flight and hence a reduced environmental impact again on a per-flight basis.

Air traffic flight procedures such as Advanced Continuous Descent Approaches (ACDA) also offer improved fuel consumption, reduced emissions and reduced overall approach time.[13] Fuel savings can also be achieved through the operational optimisation of aircraft operations. These include reducing the operational weight of the aircraft through the use of fly-by-wire technology for control systems (as opposed to heavier conventional systems), improved taxiing and optimising the aircraft speed.

The ACARE targets of achieving a 10% efficiency improvement through operations is, again, considered challenging, particularly in relation to air traffic management. In Europe, airspace is currently divided into segments, thereby causing inefficiencies in flight routes. The desire for a 'one European sky' for air traffic control is being widely pushed by the industry, but is currently set against a number of political and technical barriers. If these are overcome, fuel efficiency gains are certainly possible.

However, if as a result of the air traffic management improvements it is possible to make more flight opportunities available, an increase in the growth rate could, in only a short space of time, negate any efficiency gains made. This is known as the 'rebound effect'.[14]

Rapid Fleet Renewal

Increasing the rate at which new aircraft enter the fleet could improve the average fuel efficiency of the fleet depending on how quickly the older aircraft are retired. For airlines, fleet age makes a big difference to the fuel efficiency per passenger-km. Again, this is why the low-cost carriers such as Easyjet can claim to have the 'greenest' fleets. With aircraft generally less than three years old, compared with an average aircraft age of eight years for legacy airlines, the yearly incremental improvements in aircraft and engine design lead to lower fuel consumed per passenger-km for a low-cost carrier operating the same aircraft as a legacy airline.

At a recent Tyndall aviation industry workshop, it was argued by one of the low-cost carriers that they are putting additional pressure on manufacturers not only to accelerate efficiency improvements, but also to bring more efficient aircraft on stream more quickly. It was further argued that as low-cost carriers run more cycles per day than legacy or charter airlines, it is now inappropriate to measure aircraft age in years. Rather, the number of cycles flown should be a new measure. This viewpoint would suggest that aircraft renewal would, in terms of time, occur more rapidly than previously, thereby increasing the rate at which aircraft are retired and manufactured. The impact on overall fleet efficiency is difficult to deduce and will not be known for some time. If, for example, older aircraft are retired more quickly, then the older less efficient aircraft will be taken out of the fleet more rapidly, and overall the fleet efficiency may improve at faster rates than previously.

If, on the other hand, a buoyant secondhand market is stimulated within industrialising countries, fleet fuel efficiency may remain at levels similar to those achieved over the previous ten years—some 1 to 2% improvement per passenger-km each year. It is clear, however, that in many respects, the low-cost model in Europe is having an important impact in modifying traditional practices within aircraft manufacture and operations.

CONTRAILS AND CIRRUS CLOUDS

This chapter has primarily considered CO_2 and NO_x emissions as opposed to those emissions that lead to the formation of contrails and cirrus clouds—water vapour, sulphate and soot particles. Contrails are formed by particles from the engine exhaust acting as condensation nuclei. Forming contrails requires a cold, ice-supersaturated air mass to form. Given that contrails are only formed under very particular atmospheric conditions, operational measures to reduce the frequency of their formation may be most appropriate. In fact, contrail avoidance may be possible through relatively small changes in flight level, due to the shallowness of ice-supersaturation layers.[15] Some studies have considered the implications of substantial reductions in flight altitude, in order to reach an altitude at which the chance of contrail formation is reduced by 50%.[16] However, Mannstein et al. argue that a more efficient approach would be to equip aircraft with hygrometers that can detect individual supersaturated air layers and allow aircraft to fly up or down to an area of dryer air, where a contrail will not form. According to their study, a change in altitude of 2000 feet is sufficient to reduce the chance of contrails associated with a particular ice-supersaturation layer by 50%, and the required altitude change may in fact be less. If this is the case, an aircraft-based hygrometer and contrail-detector may be able to indicate this to the pilot.

If technically possible and if not requiring an excessive number of altitude changes, this flexible flight approach would have the advantage of achieving contrail avoidance without incurring the significant fuel penalty of substantially decreased altitude. In relation to the latter, it is important that the short-term and regional warming effects of contrails, cirrus and NO_x are not reduced at the expense of increasing the long-term warming (100-year duration) effects of CO_2, as a result of flight through lower, denser air. In short, if iterative, small reductions in altitude can substantially avoid contrails, this is likely to be a positive measure. A more general and significant lowering of flight levels, however, could have perverse effects in terms of mitigating climate change.

SUMMARY

The aviation sector has undergone a recent refocus in relation to environmental concerns. Although continuing to address local noise and air quality issues, scientific, political, media and public attention on climate change, coupled with security fears and rising oil prices, have combined to place new pressures on the aviation industry to face, and attempt to mitigate, its climate change emissions.

Aviation fuel use is currently directly proportional to CO_2 emissions; therefore where an airline's fuel costs are a significant proportion of their overall costs, driving down fuel consumed will clearly be to their financial advantage. Within the current era of public concern for climate change, promoting lower CO_2 emissions per passenger can be used as a selling point. This particularly applies to the low-cost carriers, where fuel is a much higher proportion of their operating costs than for legacy airlines. Even prior to the low-cost model, there have been significant improvements in fuel efficiency of some 70% over the past forty years, through improvements in airframe design, engine technology and rising load factors. More than half of this improvement has come from advances in engine technology.[11] Such improvements have resulted in an annual compound fuel efficiency gain of 1.14% in terms of seat-km per kg of fuel consumed or about 3% in terms of fuel consumed per passenger-km. However, given the very rapid improvements that were made in the early years, the current rate of improvement is somewhat lower than this long-term increase.

How fuel efficiency will improve over the coming decades remains uncertain. As presented within this chapter, there are a plethora of technical and operational opportunities for the aviation sector to improve on the fuel efficiency gains seen in recent years. The most likely contenders for the short- to medium-term appear to be the range of operational modifications to air traffic control, aircraft acquisition and aircraft configuration, coupled with the use of lighter materials in aircraft design and perhaps open-rotor engines. In the longer term, new designs for larger aircraft with, for example, a blended wing-body, will likely offer the prospect of step-change benefits by modifying some of the design constraints attached to today's large conventional aircraft.

Nevertheless, given the long lifetimes of current aircraft (up to forty years when considering the conventional business model), and the institutional barriers to change such as airport configuration and negotiations over, for example, the 'One European Sky' concept, the fuel efficiency of the whole fleet is likely to improve slowly. Taking a European rather than global view, slow rates of fleet renewal in many European nations have traditionally led to efficiency improvements per passenger-km over the previous twenty years of around 1–2% per year. Therefore, without continued pressure on the industry to mitigate CO_2, similar rates may continue. It should be noted, that during the writing of this book, motivation to mitigate emissions within the industry has significantly increased. If this continues to be the case, perhaps fuel efficiency per passenger-km of the order of 2 to 3% per year or even higher may again be possible in the coming decades. However, as will be illustrated in the next chapter, only if fuel efficiency per passenger-km (or carbon intensity if alternative fuels are considered) exceeds rates of growth in passenger-km, will emissions from the industry remain constant or fall. Unfortunately from a climate change perspective, with passenger-kms flown increasing rapidly (5.7% during the past twenty years for the EU),[17] stabilising and ultimately reducing aviation emissions currently appears highly unlikely within the coming decades.

5 Climate and Aviation Policy

INTRODUCTION

The previous chapter explored some of the opportunities for aviation to improve its fuel efficiency and carbon intensity per passenger-km. However, the speed at which the various opportunities are exploited and the overall consequence for aviation emissions, in absolute terms rather than in relative terms, relies not only on engineering and managerial expertise, but is intrinsically linked to the political and social context and legislative framework. For example, policies to restrict the most polluting aircraft landing in the UK may do little to drive innovation towards lower-emission aircraft if the wider EU does not take similar action. Similarly, concerns over flying within a blended wing-body aircraft founded on, for example, perceptions of safety, could scupper desires to make a step change in airframe design. Both now and in the future, political, social and cultural structures and drivers on a variety of spatial scales will be instrumental in shaping how the aviation industry develops. This chapter explores some of the most important policies and drivers visibly altering the industry, with the intention of grounding a number of the scenario assumptions presented in Chapter 6. In addition, the chapter provides a useful reference source for the current aviation climate debate. The chapter by no means aims to discuss all the drivers behind growing aviation demand or consumption, or indeed to provide a complete understanding of the complex interactions between the various stakeholders within the aviation system. Research investigating broader social and political dimensions of aviation is ongoing within Tyndall Manchester; findings from which will be made available on the Tyndall Manchester web site. However, some of the issues presented are informed by early outputs from this research.

GLOBAL POLICIES AND DRIVERS

By its nature, aviation is an international industry, and as such is subject to overarching international agreements. These agreements are inherently time-consuming and difficult to reform, as negotiations are often lengthy, with representatives naturally aiming to protect the interests of their nation or state. The aviation sector is not alone in this sense, with the majority of international negotiations being similarly drawn-out; as gaining all-party consensus hampers opportunities to respond to current concerns. Addressing climate change is arguably the most challenging issue the modern world has to face collectively. Urgent and comprehensive action is required on an unprecedented scale to prevent global mean surface temperatures from exceeding the level likely to bring about 'dangerous climate change' (see Chapter 2). The question therefore arises as to how the international community is to respond to this challenge and where the aviation sector fits within this response.

The Kyoto Protocol

In 1992, at the Earth Summit in Rio de Janeiro, the UNFCCC treaty[1] was agreed, with the aim of reducing emissions of greenhouse gases to combat global warming. In its original form, it set no mandatory limits on emissions of individual nations and contained no enforcement provisions. As an update to the treaty, the Kyoto Protocol was drawn up and opened for signature in 1997. It finally entered into force in February 2005 when fifty-five parties signed the agreement. Unlike the original treaty, the Kyoto Protocol, in its first phase, sets targets for Annex 1 (industrialised) countries on aggregate to reduce their greenhouse gas emissions by 5.2% below a 1990 baseline by 2008–2012. As a consequence, nations have agreed to differing commitments, with, for example, nations within the EU aiming for an 8% cut and some other nations even allowed increases, such as Iceland.

The protocol authorises a 'cap and trade' system, and provides additional 'flexible mechanisms' whereby nations can buy emissions rights from other members, encouraging the most economically efficient method for emission reduction. The principal mechanisms are:

- Clean Development Mechanism (CDM): Annex 1 nations can either buy emissions rights or fund projects claimed to reduce emissions from non-Annex 1 nations.
- Joint Implementation (JI): Emissions allowances are traded between Annex 1 nations.

In terms of monitoring and assessment, each nation is required to submit a national inventory of greenhouse gas emissions on an annual basis. This inventory covers emissions from energy use, industrial processes and land-use changes, released within national boundaries. The UNFCCC provide guidance on methods for measuring emissions. Emissions from international aviation and shipping are not required to be submitted for the emissions inventory, but are recorded as a memo under 'bunker fuels'.

Emission-reduction commitments under Kyoto therefore omit greenhouse gas emissions produced by international aviation and shipping, although they do include domestic aviation and inland and coastal waterway shipping emissions. This is not to say the Kyoto Protocol completely ignores emissions from international aviation and shipping, but in the absence of an agreed process for apportioning emissions for international journeys, it requests emissions be addressed through those international governing bodies responsible for aviation and shipping, namely, the ICAO and the International Maritime Organisation (IMO). The text within the Protocol reads:

> The Parties included in Annex I shall pursue limitation or reduction of emissions of greenhouse gases not controlled by the Montreal Protocol from aviation and marine bunker fuels, working through the International Civil Aviation Organisation and the International Maritime Organisation respectively.

The next section explores some of the possible policies and mechanisms available for use by ICAO, and the action it has taken subsequently.

International Response

Applying a fuel tax to aviation is one of the policy options frequently referred to in discussions relating to curbing aviation's rapid emissions growth. However, under an interpretation of Article 24 of the 1944 Chicago Convention, there can be no tax levied on fuel for aviation, either as fuel duty or value-added tax.[2] This interpretation has been implemented through some 4,000 bilateral agreements between nations, making renegotiation difficult and highly time-consuming—though not impossible. A tax on aircraft fuel is likely to be complicated by the difficulty of securing united global action, while unilateral action may result in 'tankering'—fuel being purchased in countries where it is not taxed, and transported to where it is needed. Note that the nominal difference between a tax and a charge is that a tax is applied to raise revenue, whereas a charge is often applied to change behaviour. In fact, taxes and charges may both raise revenue and change behaviour, but the distinction is one of intent. ICAO currently has no plans on an international scale to authorise the levying of taxes or charges by itself or through any other international body.

ICAO's current strategic objectives for 2005 to 2010 are split into six categories, two of which apply directly to climate change.[3] One is '*Environmental Protection*—minimise the adverse effect of global civil aviation on the environment' and the other is '*Efficiency*—enhance the efficiency of aviation operations'. Within the *Environmental Protection* objective, two measures for action are identified; the first is to adopt and promote new or amended measures to reduce noise, local pollution and greenhouse gas emissions. The second is to cooperate with other international bodies, and in particular the UNFCCC, to address aviation's contribution to climate change. The *Efficiency* objective relates to enhancing aviation operations by addressing issues that limit the efficiency of global civil aviation through a range of measures including:

- facilitating increased traffic through optimising the use of existing and emerging technologies
- developing guidance for States supporting the sustainable development of aviation
- assisting liberalisation of the economic regulation of international air transport
- improving the efficiency of operations through technical cooperation programmes

Contradictions between the *Environmental Protection* and *Efficiency* objectives occur if improvements in fuel efficiency are lower than increased traffic growth facilitated by liberalisation and better air navigation plans (see Chapter 6).

Most of the work in addressing noise and engine emissions is undertaken through ICAO's Committee on Aviation Environmental Protection (CAEP) and comprises members and observers from states, intergovernmental organisations and nongovernmental organisations representing aviation industry and environmental interests. Nearly ten years ago, CAEP requested the IPCC to produce a report on *Aviation and the Global Atmosphere*[4] and again for an update to this report to be published within the IPCC's Fourth Assessment Report in 2007. The original report comprehensively assessed the impacts and potential for emission mitigation from the aviation industry.

ICAO produce their own 3-yearly statement of policies and practices related to environmental protection, with the current statement discussing principally noise as opposed to climate change.[5] In ICAO's Assembly Resolution A35-5 in 2004, emission-related levies are discussed with states urged to refrain from unilateral implementation of greenhouse gas emissions charges prior to the next regular session of the Assembly, due in 2007. In the meantime, ICAO states that it has been studying the effectiveness of emission levies, but only in relation to local air quality. Finally, ICAO's 2004 resolution endorses further development of an open emissions trading system for international aviation, either as a voluntary system or within a system consistent with the UNFCCC.

In a subsequent development early in 2007, ICAO announced an agreement to propose guidance for incorporating international aviation emissions into emissions trading schemes consistent with the UNFCCC process. The scheme will be aimed at CO_2 alone. Further illustrating the global aviation industry's concern for its carbon emissions is the publication of ICAO's first Environmental Report (2007). This report includes a series of adverts endorsing continued growth in flying, suggesting ICAO believe either continued high growth in aviation is consistent with global emissions being curtailed, or accept that aviation's emissions will form an increasing proportion of total emissions. The absence of policies in place to actively address aviation emissions has led the EU to be highly critical of ICAO's role in tackling the issue, stating in October 2007:[6]

> Regrettably, it has become clear to us at this 36th Assembly that, ten years after having been requested by the UNFCCC to take action to limit or reduce emissions, it has not been possible for ICAO to agree on essential elements of this comprehensive approach. In particular, the programme put forward for agreement at this Assembly is unambitious, piecemeal and lacking in credibility on market-based measures (both greenhouse gas emissions charges and emissions trading).

EU POLICIES AND DRIVERS

As international action to combat aviation emissions has been limited, with no binding targets, emissions trading, taxes or levies in place in the ten years since the Kyoto process began, the EU has decided to respond unilaterally. The drive for such a response was made clear in a press release from the EU at the end of September 2007, where the EU Commissioner Stavros Dimas stated:

> In order to fight climate change, all sectors must contribute in a fair way, including aviation, whose emissions are increasing very rapidly. It is a great pity that ICAO has not been able to reach an agreement on the way forward. The EU has set up an ambitious and comprehensive emissions trading system and is in the process of agreeing legislation that would extend it to aviation emissions. This process must continue without delay.

Clearly, and despite complications relating to the international nature of the emissions from aviation, the EU Commission wishes to ensure all sectors are playing their fair role in mitigating emissions. As discussed in Chapter 3, the EU has set a non-binding target for greenhouse gas emissions to be reduced by between 60 to 80% by 2050; targets that do not explicitly exclude international aviation and shipping emissions, exclusions made explicit in the UK's CO_2 targets. In tackling emissions from aviation, the EU's central policy is to pass legislation to bring aviation into the EU ETS. The details of how and when to include aviation are currently subject to negotiation and discussion.

EU Emissions Trading Scheme

The EU ETS began operation in January 2005, with the first phase of the scheme set to run until the end of 2007. The scheme initially involved some 12,000 installations covering energy activities that exceeded 20MW, as well as a number of process emission activities, together covering around 45%[7] of the EU's domestic CO_2 emissions. The second and expanded phase of the EU ETS begins in 2008, and, in recognising the growing issue of emissions generated by the aviation industry, the EU has recently announced it wishes to include aviation within the scheme shortly after, with the actual date still under discussion.

The aim of including aviation within the scheme is to internalise some of the costs of the environmental impact of the aviation sector. However, a number of issues are yet to be fully understood or decided, in particular; the impact of including aviation in the ETS, the level of emissions rights to be purchased by the aviation industry given current emission growth rates and the overall level of the EU ETS cap. Issues central to including flights within the EU ETS have been examined in detail by CE Delft,[8] with their report including how to take account of the additional warming effects of high-altitude emissions and the range of flights to be included or excluded. Clearly, the greater the number of flights covered by the scheme, the more likely it will have a meaningful impact in terms of climate change. Flight coverage options proposed by CE Delft were:

- emissions from all flights departing the EU
- emissions from all intra-EU flights
- the emissions from flights passing through EU airspace

The option likely to have the biggest impact on reducing greenhouse gas emissions is the inclusion of all flights departing the EU, whilst other nations take responsibility for the emissions of flights departing from their territories. More specific recommendations of CE Delft are to exclude:

- flights operating on Visual Flight Rules (VFR flights)
- flights under military Air Traffic Control flight rules and flights with a military flight purpose
- flights operated by aircraft with less than 8,618 kg maximum takeoff weight

Furthermore, CE Delft recommended that to avoid leakage in the form of traffic shifting to low-traffic airports, only very small airline operators or airports should be excluded. For these operators, the threshold of less than one flight per day was considered reasonable, given that this reduces the number of trading entities (airlines and airports) by 55%[8] and omits 'only' 3% of traffic and emissions. The possibility of using some annual per-aircraft emissions threshold to capture any shift to smaller aircraft and air taxi operations is also under discussion.

To account for the additional warming effects, an accompanying NO_x charge is presented in the CE Delft report as a possibility, but this still omits the warming arising from water vapour, soot and sulphur emissions as described in Chapter 2. It should be noted that such a charge would only reduce warming from NO_x significantly if it were high with respect to the charge levied on CO_2. Reductions are more likely to be driven by a high CO_2 charge or a stringent EU ETS carbon cap, particularly given the close association of NO_x and CO_2 with fuel burn. Policy instruments to encourage the avoidance of contrails have rarely been discussed, but the suggestion of Mannstein et al. in 2005,[9] whereby aircraft use a hygrometer (Chapter 4) and perhaps an atmospheric forecast to alter flight path, merits further analysis. In due course it could, for example, be mandatory for pilots to modify their altitude upon producing a contrail if the available information suggests that such a change in altitude would take them to drier air (and if it was safe to do so).

Since the report by CE Delft, including aviation within the EU ETS has generated considerable interest from academics, politicians, economists and industry alike. Many airlines and airport operators within the EU are broadly supportive of including aviation within the EU ETS, although they differ in their views as to which flights should or should not be included. British Airways, for example, are keen for aviation to be included as soon as possible, but for the scheme to cover just intra-EU flights in the first instance, with a broader scheme brought in eventually.[10] Some of the low-cost airlines, on the other hand, are keen to see the scheme covering all flights from the outset, perhaps seeing themselves at a disadvantage compared with airlines whose complete portfolio of flights may not be covered from the beginning.

Since CE Delft outlined their three options for including aviation within the EU ETS, the EU Commission have proposed a further option—a scheme covering all arrivals and departures to and from the EU, starting with intra-EU flights in 2011 and all flights in 2012. This proposal is in response to an absence of a comprehensive global emissions trading scheme in which aviation is included. By choosing to include all flights, the EU is essentially taking responsibility for other regions' aviation emissions, thereby demonstrating a clear leadership role in addressing the issue. Should a comprehensive scheme covering aviation be eventually established, it is likely the EU will adjust its scheme to cover a 'fair' proportion of the flights arriving and departing the EU.

Despite the considerable administrative burden involved in including the aviation industry within the EU ETS, the moves currently being made are seen by many as a key step towards a consistent and sector-wide effort to combat climate change. However, in the period prior to the scheme commencing, the EU's emissions from aviation are expected to grow significantly. Whether or not the aviation sector will be required to buy permits to account for the emissions released between now and the scheduled start date is currently under discussion and, depending on the price of carbon, may be key to the scheme's success.[11]

A recent recommendation from the EU Parliament Environment Committee[12] suggests commencing the scheme, including all flights, from 2010 and with a baseline CO_2 amount of 75% of the 2004 to 2006 level. In other words, the aviation industry would be required to purchase any CO_2 emitted above this level when it joins the scheme. By rough calculation,[13] this would be equivalent to a baseline of around 46MtC, given emissions for all departures and arrivals are estimated at 61MtC for 2005. This is lower than the 2000 baseline estimate of 54MtC. However, in November 2007, the EU Commission backed proposals for the cap to be set at 90% of 2004–2006 levels. The amount to be purchased will therefore depend on the final agreed cap, and aviation's growth in the meantime. This is explored more fully in a recent Tyndall report (Anderson et al. 2007).

The success of the EU ETS in addressing climate change is highly dependent on the overall emissions cap (the cap for all sectors or installations within the scheme), and the rate at which the cap is ratcheted down year-on-year. Given the uncertainties surrounding how to include the aviation sector within the EU ETS, if and how the overall cap will be adjusted to allow aviation's entry into the scheme is even more unclear. For example, will the cap be raised when the aviation industry is included, and if so, by how much? Will there be a reduction in the cap year-on-year, and if so, what will the rate of change be? Table 5.1 qualitatively illustrates these issues and their likely impact on the cost of carbon allowances. However, the categorisation assumes a low or zero use of the CDM or JI. CDM in particular could significantly undermine incentives for domestic emission reductions.

Table 5.1 How the Emissions Cap within the EU ETS May Affect the Price of Carbon Permits Assuming No Use of CDM or JI

	No Emission Cap Reduction	*Small Emission Cap Reduction*	*Large Emission Cap Reduction*
Extra allowance for aviation (at a level similar to the total value of aviation emissions)	No change in permit price	Small increase in permit price	Moderate increase in permit price
No extra allowance for aviation	Small increase in permit price	Moderate increase in permit price	Large increase in permit price

Taxes and Charges

Another option that could be used in conjunction, and indeed prior to, aviation being included within the EU ETS, is to apply a Europe-wide emissions charge, where airports would be required to levy the charge on all aircraft, passenger or freight, taking off from or landing at European airports. The charge could differentiate between aircraft types and loads and between the distances travelled over Europe, to reflect estimated emissions. As ticket prices would likely rise, this would reduce growth in demand, particularly in the leisure sector, which has, a least for marginal price rises, a higher elasticity of demand than the business sector. How to choose exactly which emissions to apply a charge to, and indeed how this could be squared with the EU ETS is not clear. Nevertheless, in principal, it may be appropriate for a sector to have its emissions subject to multiple environmental policy instruments, if they collectively achieve an end to which society agrees.[14] In this case, the objective is to permit a reasonable level of air transport whilst protecting the integrity of the climate system. Whether to apply an emission charge to different gases or emissions is arguably a more challenging decision. In 1999, the European Commission Communication *Air Transport and the Environment—Towards Meeting the Challenges of Sustainable Development*[15] stated that the European Commission would be carrying out preparatory work with a view to possibly introducing proposals to establish economic incentives to mitigate the greenhouse gas emissions of air transport in Europe. The EC commissioned consultants to investigate the feasibility and effects of an aircraft emissions charge for intra-European flights, including a comparative study of the environmental benefits of a Performance Standard Incentive (PSI). This would financially reward the best-performing aircraft, in terms of CO_2 and NO_x emissions, with revenue obtained via penalties imposed on the poorer performers.

With mid-range charge values of €30 per tonne of CO_2 and €3.6 per kg of NO_x emitted, the study[16] forecast that an environmental charge would reduce aviation CO_2 emissions in EU airspace by about 10 million tonnes (9%) in 2010, which is the time horizon of the study, compared to the level that emissions would otherwise reach. Half of this reduction would be as a result of technical and operational measures by airlines and the other half as a result of reduced air transport demand. Their study suggests that there would be a rise in average ticket prices of roughly €3 to €5 for short one-way flights (500km) and €10 to €16 for long flights (6,000km).[16] An alternative mechanism, the PSI, was estimated to reduce aviation CO_2 emissions in EU airspace by almost 6 million tonnes (5%) in 2010 compared with a 'business as usual' situation, according to Wit et al. This would accrue almost entirely from technical and operational measures by airlines. The impact on ticket prices depends very much on the precise definition of the 'performance standard'. The PSI does not place a net financial burden on the industry as a whole. By its very nature, though, the introduction of the PSI may mean that some market segments benefit and others suffer. In relation to an emissions charge, it is highly likely that it would need to be considerably higher than levels modelled by Wit et al., if aviation were required to make the same CO_2 emissions reductions as other sectors in aiming for the 2°C threshold.

Tax on aviation fuel is another area being explored currently by the EU. Although it is not possible to tax aviation fuel on international flights under the 1944 Chicago Agreement, domestic aviation fuel can be taxed. Furthermore, by renegotiating bilateral agreements between nations, fuel tax can be imposed. It is understood that a number of such agreements are already in place; therefore the possibility of a fuel tax in conjunction with the EU ETS inclusion is now more likely than previously. Indeed, in 2006, the EU Environment Committee urged the Commission to make plans for putting a fuel tax on domestic and intra-EU flights to redress the imbalances between aviation and other sectors.

High-Speed Train as a Substitute for Short-Haul Flights

Tackling the scale and urgency of climate change demands not only addressing growing aviation emissions through policy and regulation aimed specifically at the aviation sector, but also in establishing viable alternatives to encourage a modal shift from air travel to rail, for example. Short-haul flights are particularly fuel intensive, given takeoff and landing use far more fuel per passenger-km than does the cruise phase. Rail travel is significantly lower in terms of CO_2 per passenger-km than flying short-haul, and can have almost zero emissions if travelled by high-speed train powered by low-carbon electricity (as is the situation in France).

On long-distance routes (over 800 km or 500 miles), air transport tends to dominate the transport market, but even here there are opportunities for complementarity in terms of providing feeder traffic for long-distance flights by high-speed rail.[17] In Europe, long continental and intercontinental flights could be fed by high-speed rail, instead of equivalent short-haul air. This kind of multimodal interrelationship requires, for example, multimodal integration, including through-ticketing, timetable integration and common baggage processing. Experience and research to date show that European high-speed rail has mostly affected air travel on the origin/destination passenger market corridors of between 400 and 800 km (250–500 miles). It has been possible to achieve this because high-speed rail has been able to run services at speeds of up to 300 km/h, which together with a competitive departure frequency, has produced total station-to-station travel time of between 2 to 4.5 hours.[17] Other policies discussed in this chapter would likely serve to increase the substitution of high-speed rail for air, at least for short- and medium-haul journeys, the benefits of which would be further increased with a greater use of electricity supplied from lower- or zero-carbon sources.

UK POLICIES AND DRIVERS

In the UK, the Government White Paper, *A New Deal for Transport—Better for Everyone*,[18] not only set a framework for future transport policy, but also expressed a commitment to a 'sustainable' transport system. The Department of Environment, Transport and the Regions (DETR) defined such a system as one that supports employment and a strong economy, increases prosperity, addresses social exclusion, does not damage human health and provides a better quality of life for all now and in the future. In July 2002, the UK Department for Transport (DfT) released its consultations on Regional Air Services, which detailed specific regional options for where and how airport growth might be accommodated. The mid-range RASCO (Regional Air Services Co-ordination) scenario assumed a near trebling of UK air passenger demand by 2030. In March 2003, the DfT and HM Treasury issued a consultative policy document, *Aviation and the Environment: Using Economic Instruments*.[19] This was intended to support discussion with stakeholders regarding economic instruments for encouraging the industry to take account of, and where appropriate reduce, its contribution to global warming, local air and noise pollution.

Then, in December 2003, the DfT issued the Aviation White Paper—*The Future of Air Transport*.[20] This was nominally intended to balance increasing demand for air travel with environmental protection, taking the view that the UK's economy depends on air travel, with many businesses, in both manufacturing and service industries, relying heavily on this mode of transport. Airfreight was noted to have doubled in the preceding ten years, with 200,000 people now employed in the aviation industry, an estimate of three times as many jobs supported indirectly, and visitors arriving by air considered crucial to UK tourism. According to the DfT, all of the above put pressure on airports, some of which are at, or fast approaching, capacity. Therefore, the White Paper set out what the DfT considered to be a measured and balanced approach providing a strategic framework for the development of air travel over the next thirty years. Demand projections for passenger numbers within the White Paper were broadly consistent with the mid-range RASCO scenario referred to earlier, with forecasts ranging from 400 to 600 million passengers passing through UK airports in 2030 compared with 200 million in 2003. In relation to infrastructure, the White Paper set out plans for a new runway at Birmingham, Edinburgh, Stansted and Heathrow airports, plus new terminals, apron and runway extensions throughout the UK.

In the White Paper's mid-level forecast, aviation passenger growth in the UK—a country with a relatively mature aviation industry—averages 3.3% per year between 2000 and 2030. This figure is based on a growth of around 3.8% per year in terms of passenger numbers until 2020, then a further growth of 1.8% per year from 2020 to 2030. DfT's high-level forecast shows an average growth of 4% per year up to 2030—4.5% between

2000 and 2020, and around 2.7% from 2020 to 2030. Historically, growth in passenger numbers at UK airports has averaged around 5.8% from 1973 to 2003,[21] substantially higher than DfT's future projections. Moreover, the Eurostat dataset[22] suggests that the current rate of growth in the UK is actually 6.4%, based on the trend between 1993 and 2001 (eliminating the short-term effects following the events of September 11), again higher than the 3% to 4% assumption used in the White Paper.

In the same year of publication as the Aviation White Paper, the UK Government produced its 2003 Energy White Paper—*Our Energy Future—Creating a Low Carbon Economy*[23] in which it accepts the RCEP recommendation of a reduction in CO_2 emissions of 60% by 2050 (Figure 3.5 in Chapter 3). Although not explicitly including aviation within this aspirational target, by basing their aim on the RCEP's conclusions, which were reached following an analysis considering the 2°C threshold, the DTI implicitly accepted all sectoral emissions must be included. The 2°C temperature threshold was linked to a particular global atmospheric CO_2 concentration as explained in Chapter 3; therefore if individual nations choose to omit sectors, clearly the global CO_2 concentration desired will not be attained. It is the conflict between the UK Government's transport and climate change objectives that originally generated the interest of the Tyndall Centre in addressing the problem.[24]

Within the 2003 Aviation White Paper, the DfT not only include forecasts of passenger growth, but in addition provide emission estimates for a range of futures. However, these forecasts were updated shortly afterwards as they were considered to underestimate fuel efficiency improvements from engines and airframe changes. Updated forecasts were then published within documentation accompanying the White Paper[25] and are presented in Table 2.1 in Chapter 2. These emission forecasts are based on the high-capacity cases for the UK,[19] assuming new runways at Heathrow, Gatwick, Stansted, Manchester, Birmingham and Edinburgh and varied assumptions relating to fuel efficiency. The forecasts are illustrated graphically in Figure 5.1 and summarised in the following.

- '*Worst case*'—assumes limited fuel efficiency improvements, limited fleet renewal and no economic instruments. The case is based on the 'high-capacity' case but assumes three, rather than two, additional runways in the South East of England, and unconstrained capacity in the regions.
- '*Central case*'—based on the 'high-capacity' figures, but incorporating fuel efficiency improvements envisaged by the IPCC[4] and by ACARE.[26]
- '*Best case*'—based on the 'high-capacity' figures but taking account of the effect of economic instruments (e.g., a fuel charge or carbon price), to produce an additional 10% fuel efficiency saving from 2020 onwards, with half of that occurring by 2010.

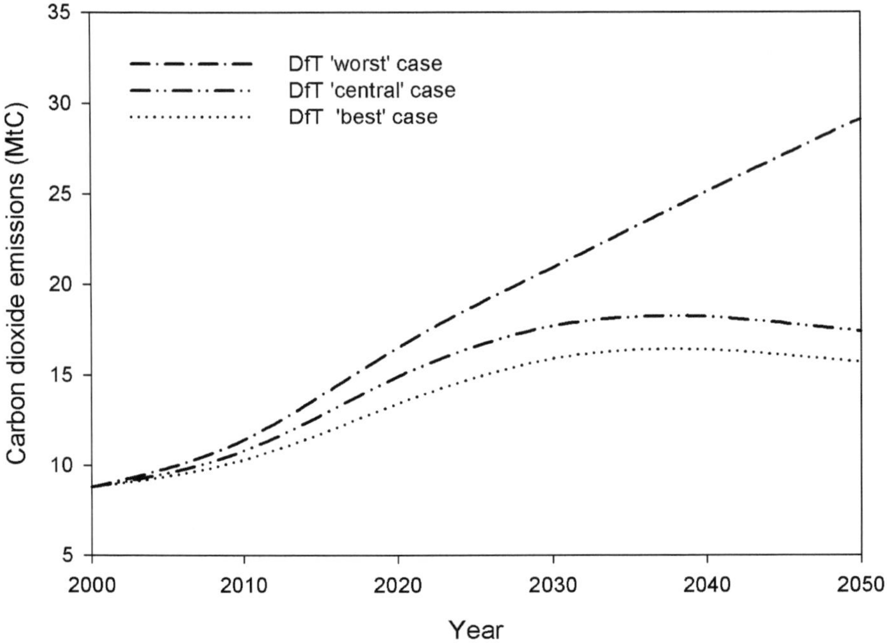

Figure 5.1 DfT aviation CO_2 2003 emission forecasts. (From Aviation and Global Warming, 2003.)

Apart from the 'worst' case forecast, emissions increases due to growth within the aviation industry will be offset by efficiency improvements from 2030 onwards according to the DfT. To understand the scale of the forecast emission increases in relation to the UK's climate change goals, the emissions from aviation are now compared with the emission-reduction pathways presented in Chapter 3. Firstly, the Government's aviation forecasts are compared with the original Contraction & Convergence profile used to deduce the 60% carbon-reduction target (first illustrated in Chapter 3, Figure 3.5). In addition, the corresponding profile for the 450 ppmv CO_2 level is presented; 450 ppmv CO_2 offers a much greater probability of not exceeding the 2°C threshold than the 550 ppmv level. The results are illustrated in Figure 5.2.

By comparing the aviation emission forecasts with the carbon-reduction pathways, it is clear that if the UK is to successfully reduce its emissions in line with the 60% carbon-reduction target, emissions from aviation will increasingly contribute a growing proportion of those emissions. By 2050, the DfT's 'worst' case forecasts equates to some 50% of the UK's total carbon profile (assuming the 60% target); with the 'central' and 'best' cases assuming around 25% of the total carbon allowance. If, on the other hand, a national carbon profile more consistent with the Government's 2°C temperature target is considered, the aviation industry contributes almost

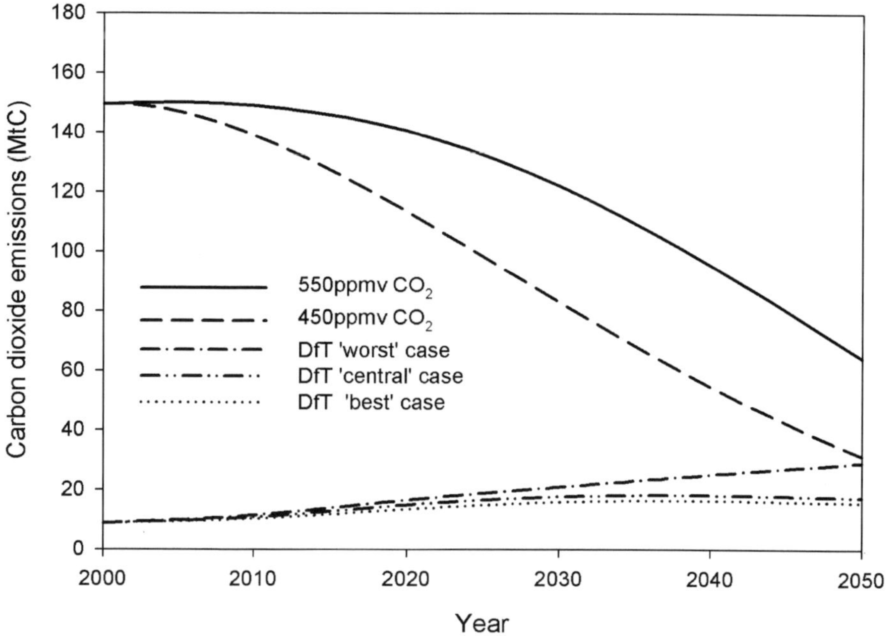

Figure 5.2 450 ppmv and 550 ppmv CO_2 emission pathways for the UK compared with the DfT's aviation emission forecasts.

100% of emissions by 2050 in the 'worst' case, and around 50% in the other cases.

However, given that this national carbon profile assumes UK emissions begin reducing immediately, it is important to make comparisons with more realistic profiles in which reductions are delayed. Such profiles were illustrated in Chapter 3, Figure 3.6, and are based on a comprehensive global emissions inventory in which international bunker fuels are included. If the UK Government's aviation forecasts are compared with these profiles (Figure 5.3), the message is more stark, with all three of the DfT's aviation scenarios absorbing or exceeding the UK's total carbon allowance by 2050.

The aviation industry currently accounts for over 6% of the UK's CO_2 emissions. The shift to much higher proportions has significant consequences for other economic sectors, as they will require radical decarbonisation to compensate for the aviation industry's emissions. Allocating such a large proportion of emissions to one particular sector will inevitably have significant policy implications for society as a whole. This issue will be discussed further in Chapter 7.

EU Emissions Trading Scheme

Following the publication of the Aviation White Paper, the UK House of Commons Environmental Audit Committee vigorously debated projected aviation

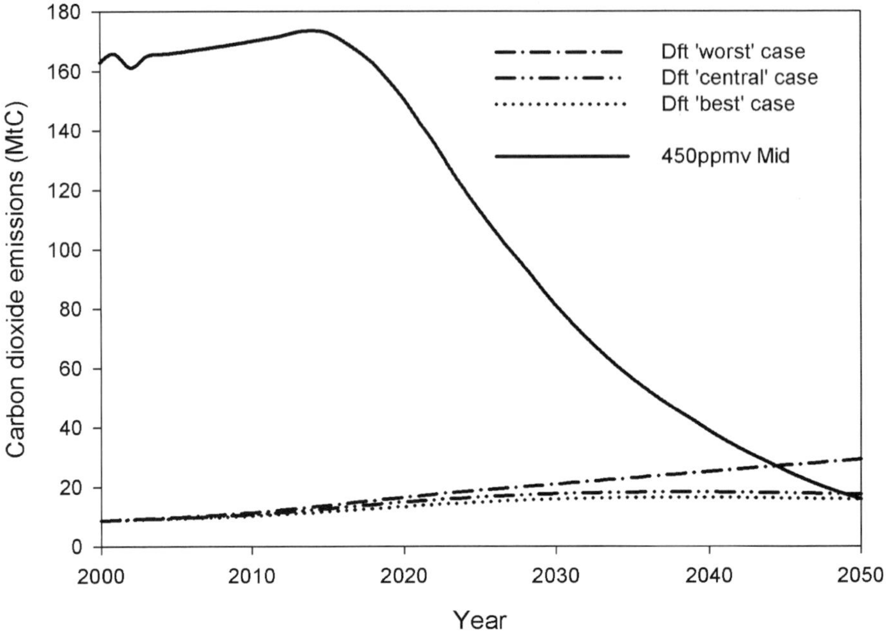

Figure 5.3 A comparison of the DfT's aviation CO_2 forecasts with the mid-range UK carbon budget taken from Figure 3.6 in Chapter 3.

growth and its impacts with the DfT. The Environmental Audit Committee's concern about the impact of aviation emissions on the UK's long term CO_2 reduction target was echoed by the House of Lords' EU subcommittee on environment and agriculture in November 2004,[27] which recommended incorporating the full climate-forcing effects of intra-EU aviation emissions into the EU ETS at the earliest possible opportunity. Furthermore, in 2007, the UK Government produced its first draft of the Climate Change Bill.[28] The Bill, in its draft form, recommends the UK makes a 60% reduction in CO_2 emissions by 2050, with a 26% to 32% reduction by 2020. However, the Bill explicitly excludes aviation, stressing this sector will be tackled through incorporating it into the EU ETS.[29] As presented in Section 5.3, discussions relating to when and how to include aviation within the EU ETS are ongoing.

The UK Government has been very supportive of including aviation within the EU ETS, as have other political parties within the UK. However, the impact for the UK of including aviation within the scheme is still to be determined. One issue of interest is the very large proportion of emissions generated by aviation in comparison with the CO_2 emissions from those UK sectors already within the scheme. Given that the EU Commission proposes to include all arrivals and departures within the EU ETS, the UK's aviation industry could account for a very significant proportion of the UK's allocation. In 2006, the UK's aviation industry, domestic CO_2 plus 50% of international flights, amounted to around 11MtC. If departures and arrivals are taken into account, this figure could be closer to 18MtC.[30] Therefore, even if emissions from this sector remain static until aviation's inclusion, and if all flights are to be within the scheme, this represents around 28% of the UK's total allocation, assuming no extension of emissions sectors within the scheme. On the other hand, if the aviation industry is only required to purchase the difference between 90% of 2004 to 2006 levels, plus any additional emissions when entering the scheme, this proportion will be much reduced. Recent Tyndall research suggests that if the carbon price remains low (i.e., below €50 per tonne), and aviation is required to purchase only a small percentage of its emissions, then the impact of the scheme is likely to be marginal (Anderson et al. 2007).

Air Passenger Duty

One other policy being applied to the UK's aviation sector by the current Government is Air Passenger Duty (APD). APD came into affect in 1994 and is a duty of excise levied on the carriage, from a UK airport, of chargeable passengers on chargeable aircraft. Until recently, APD stood at £10 for specified European destinations, £40 for all other destinations at standard rates, and £5 and £20 respectively for reduced rates. The reduced rates apply to economy class of travel. More recently, and in an apparent reaction to tackling growing emissions from the aviation sector, the current Government chose to double the APD as a way of curbing emission growth by impacting passenger demand. This move proved hugely unpopular with the aviation industry, many of whom criticised the use of such a tax for environmental reasons without ring-fencing the money to address environmental issues. It was felt by some members of the industry, interviewed during Tyndall aviation work, to be highly unfair given the industry's support and cooperation for including aviation within the EU ETS. Since the doubling of APD, the UK pre-budget report of 2007 stated the Government's intentions to modify the tax to apply to aircraft rather than passengers. Whilst there has been a mixed response to the decision from the industry, there is a broad agreement that it will benefit those carriers who have higher load factors and seat densities (see Chapter 4). The new form of tax is due to be applied from November 2008, and until then, the current form of APD remains. Whilst the detailed impact of the rise in APD tax, in terms of demand management, is unclear, it is very unlikely to significantly change either the price or economic drivers for flying.

SUMMARY

This chapter has covered some of the current global, EU and UK policies affecting the aviation industry from the perspective of climate change and emissions mitigation. The current view is that ICAO has done little to address aviation's rapidly growing emissions at the global scale, and therefore the EU has begun to take a lead role in tackling the issue by setting out proposals for including aviation within its emissions trading scheme. Whilst this course of action has drawn broad support from the aviation industry across Europe, the inclusion of all departures and arrivals has already led to talk of legal action against the plan from the United States. Although the EU's lead position in addressing the issue through emissions trading is to be commended, the impact on the rapidly growing emissions will, if the current scheme and weak caps are any measure, likely be inadequate for reconciling emissions with the 2°C aspiration.

The emission reductions required for the EU to play its fair role in addressing climate change are far more challenging than generally recognised. CO_2 emission reductions have not been evident to date, indeed CO_2 emissions continue to rise; therefore a portfolio of stringent and urgent policies and regulations will need to be implemented if emissions are to genuinely be brought in line with 2°C.

There are a variety of options available for mitigating aviation's impact on the climate. However, given the industry's relative high baseline and very rapid rise in emissions against the backdrop of stringent carbon-reduction pathways, measures to curb growth in passenger-km will undoubtedly form a necessary element of any meaningful policy portfolio. The emission-reduction pathways for both the EU and UK will require unprecedented efforts across all sectors in relation to fuel-shifting, improving technological and operational efficiency and modifying energy-consuming practices by organisations and individuals. These changes will need significant investment, time and sustained political will. Unfortunately, despite binding and nonbinding emission-reduction targets, neither the UK nor the EU are close to seeing even moderate CO_2 emission reductions year-on-year. Even if emissions do begin to reduce at the rates required from other non-international sectors, the continued growth in emissions from the aviation sector will significantly undermine, if not negate, any reductions made.

6 Comparative Assessment

INTRODUCTION

This chapter integrates the information discussed in previous chapters, namely climate change emission pathways and budgets, aviation growth, technology, operations and drivers of aviation demand, as well as the policy and regulatory framework. The emission pathways and budgets presented in Chapter 3 will be compared with aviation emission scenarios for the EU's and UK's aviation industries. The impact a growing aviation industry may have on the EU's and UK's contribution to not exceeding a 2°C temperature rise above preindustrial levels is discussed.

AVIATION EMISSION SCENARIOS FOR THE EU

During the process of drawing together research for this book, the policy framing and scientific understanding of both aviation and climate change have changed considerably. For example, the continued high levels of emissions from industrialised nations coupled with a better understanding of the importance of cumulative emissions, mean even a few years of additional emissions can change the policy response required. Furthermore, until fairly recently, the 550 ppmv CO_2 level was closely aligned at least amongst the policy community, with the 2°C temperature threshold, whereas stabilising at this level is now thought to offer very little chance of remaining within the 2°C target. This also modifies the results significantly. To illustrate how this research has developed over the last two to three years, both old and new scenarios and results are presented. The old scenarios are referred to as '*preliminary scenarios*' and the new ones are referred to as 'updated scenarios'. They are compared with the EU and UK emission pathways available at the time. It is the updated scenario, for both the EU and UK, where the current understanding of the scale of the problem lies. This should be borne in mind when considering the 'preliminary scenarios'. Through presenting the work in this way, it is hoped the reader will be able to better appreciate the importance of considering up-to-date empirical data and recent scientific developments on the climate policy framework within which the research has been conducted.

Preliminary EU Aviation Scenarios

The availability of detailed public domain data relating to the growth in emissions from the aviation industry is limited, particularly for nations other than the UK. Although detailed aviation emissions modelling can be carried out, it requires access not only to a range of origin and destination data, but also to aero-engine and route models. Such modelling would be a bottom-up approach to estimating both current and future emissions. A top-down method would be to gather data corresponding to fuel consumed, and use emission factors to estimate the associated CO_2 emissions, or similarly, use the CO_2 emissions for domestic and international aviation submitted to the UNFCCC. Gathering the emission trends over a number of years provides a grounding for how emissions may develop in the future, given a range of assumptions relating to fuel efficiency and growth. However, the quality of the data appears to vary widely from nation to nation. The aviation scenarios developed here for the EU therefore begin by using what was considered, at the time, to be the most reliable data source—passenger numbers. These particular scenarios were developed in 2005, therefore some of the data sources may have since been updated or improved.

Although it is aircraft that directly emit greenhouse gases, not passengers, passengers are obviously a key driver for aircraft traffic. According to the DfT,[1] both UK passengers flying on UK airlines and UK passengers passing through UK airports grew on average at 7% per year between 1993 and 2000, during which time fuel consumption grew on average at 6% per year.[2] This illustrates the close relationship between passenger growth rates and fuel consumption growth rates for the UK's industry. It should be borne in mind however, it is the passenger-km growth rather than simply the passenger numbers that is key to understanding emission growth. It is therefore assumed within this method that there is no significant modification to the lengths of flight being taken. If however, flights tend to increase in length, then the passenger-km figure may increase more rapidly than do passenger numbers. As this is the current trend, this preliminary method will, to some degree, underestimate emission growth.

To gather passenger numbers for the year for which the most current data was available at the time (2002), Eurostat documentation was used. In 2002, the EU had only recently incorporated the twenty-five nations. Therefore historical growth figures for passenger numbers for the original EU nations and data on the accession nations were held in different datasets. For the 'old' EU nations (Austria, Belgium, Denmark, Finland, France, Germany, Greece, Ireland, Italy, Luxembourg, Portugal, Spain, Sweden, United Kingdom and The Netherlands) data were available from 1997–2001 and 1993–1997 in two separate publications.[3] For the accession nations or the 'new' EU nations (Cyprus, Czech Republic, Estonia, Hungary, Latvia, Lithuania, Malta, Poland, Slovenia, Slovakia), growth data for 1995 to 2000 were also available from a different Eurostat source.[4] Using these sources, a comprehensive dataset for passenger growth was developed for 1993–2001 for the 'old' EU15 nations, and 1995 to 2000 for the accession nations. Due to the events of September 11, 2001, having a temporary but significant impact on the growth figures between 2001 and 2003, this dataset arguably gives a better basis for estimating future passenger growth than would a dataset that included 2001–2002.

In constructing the emission scenario from this passenger dataset the period over which current trends should be extrapolated was limited, particularly given the extremely high growth figures of some nations: for example, in Spain, passenger numbers increased at about 12% per year. Given that many of the EU15 nations have what can be considered to be relatively mature aviation industries,[5] whereas the 'new EU' nations have much younger aviation industries, and hence generally more potential for growth, two distinct time limits were placed on the extrapolation of current trends. Consequently, for the EU15 nations, 1993–2001 trends were continued until 2015. Whilst for the 'new EU' nations, 1993–2001 trends were continued until 2025.

Without deliberate policy decisions for curbing the rate of air traffic and passenger growth and in the absence of external shocks, it is assumed the industry will continue growing within the time frame of this analysis (i.e., to 2050). For the UK, the Aviation White Paper suggests, by way of its mid-level forecast (see Chapter 5), that growth in the UK—a country with a relatively mature aviation industry—will average 3.3% per year between today and 2030. DfT's high-level forecast has an average growth of 4% per year up to 2030—4.5% between 2000 and 2020, and around 2.7% from 2020 to 2030. The Eurostat dataset suggests that the recent rate of growth in the UK was 6.4%, based on the trend between 1993 and 2001.

For the post-2015 or post-2025 years, the scenario assumes the aviation sectors of all EU nations will continue to grow after they reach maturity (2015 for EU15, 2025 for 'new EU'). The growth rate used across all nations is taken as that assumed for the UK in the Aviation White Paper. While this specificity would be unlikely in practice, it is not an unreasonable assumption for present purposes, particularly given the global nature of the industry. The DfT assumes an average of a 3.3% per year increase in passenger numbers between 2000 and 2030. While it currently looks as if this value may underestimate the likely trend, in the absence of airport capacity constraints, 3.3% per year is nevertheless assumed as a conservative rate for this scenario. The justification for assuming this to be a conservative growth rate in the absence of airport capacity constraints is that:

- The UK has a relatively mature aviation industry; yet contemporary passenger number increases per year are still substantially higher than

3.3% (an ongoing annual increase of 3.3% per year is well within the bounds of possibility).

- All EU nations, other than Latvia and Malta, are showing much higher annual rates of change in passenger numbers, and 3.3% represents a significant reduction in growth from current levels.
- 3.3% is only marginally above recent levels of GDP annual growth in the UK, and the aviation industry has historically grown at levels well above GDP. A similar study also recently projects UK passenger numbers increasing at 3, 4 and 5% per year up until 2050,[6] with no explicit airport capacity constraint. Figure 6.1.

If it is assumed that the underlying structure of the aviation industry remains unchanged (i.e., routes, load factors, air-traffic management, fleet and engine efficiency) then an increase in passenger numbers would result in a proportional increase in CO_2 emissions. However, reductions in the amount of carbon emitted per passenger-km are likely to arise from a combination of load factor improvement, aircraft design, aircraft size, air transport management and engine efficiency. Fleet-wide estimates for likely future improvements are often in the range of 1 to 1.5% per year. The current ACARE targets challenge the industry to a 3.4% per year fuel efficiency per passenger-km

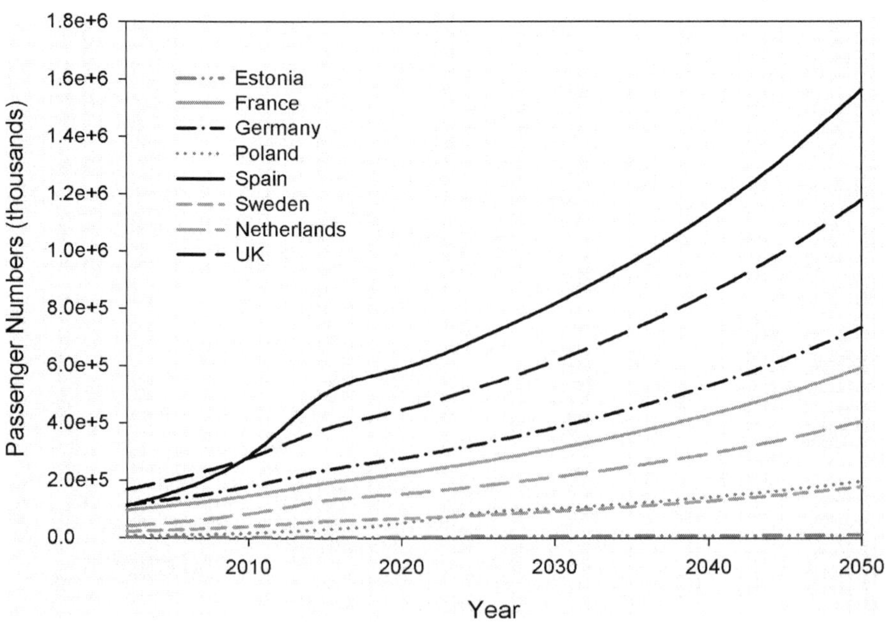

Figure 6.1 Scenario growth in passenger numbers for selected EU nations. Note: the passenger scale is a linear scale for thousands of people (i.e., 1.0e+6 = 1 million passengers × 1000).

improvement for a new plane in 2020 compared with its equivalent in 2000. However, fuel efficiency across the entire global fleet will improve at a slower rate due to the time lag involved in fleet renewal. More information on likely fuel efficiency improvements is documented in Chapter 4.

To produce corresponding aviation CO_2 emissions for all EU nations between 2002 and 2050, data from the UNFCCC are used as a baseline and increased annually by a combination of the percentage increase in passenger numbers and a fuel efficiency improvement of 1.2% per passenger-km.

The CO_2 emissions, with the inclusion of efficiency and other improvements are plotted in Figure 6.2, showing emissions from international and domestic aviation for a selected number of nations. Although growth in passenger numbers in Spain leads to a higher number of passengers than for the UK, Figure 6.2 shows the UK as having the highest emissions of all nations. This reflects the fact that, according to the UNFCCC data, aviation emissions in the UK are currently the highest in Europe by some margin, with the large hubs, such as Heathrow, adding considerably to the UK's emission burden. It may also point to a discrepancy between the UK's and Spain's emission estimation methods. The kinks in the emission scenarios presented are a function of the change in growth assumed within the scenarios occurring as a nation's aviation industry matures.

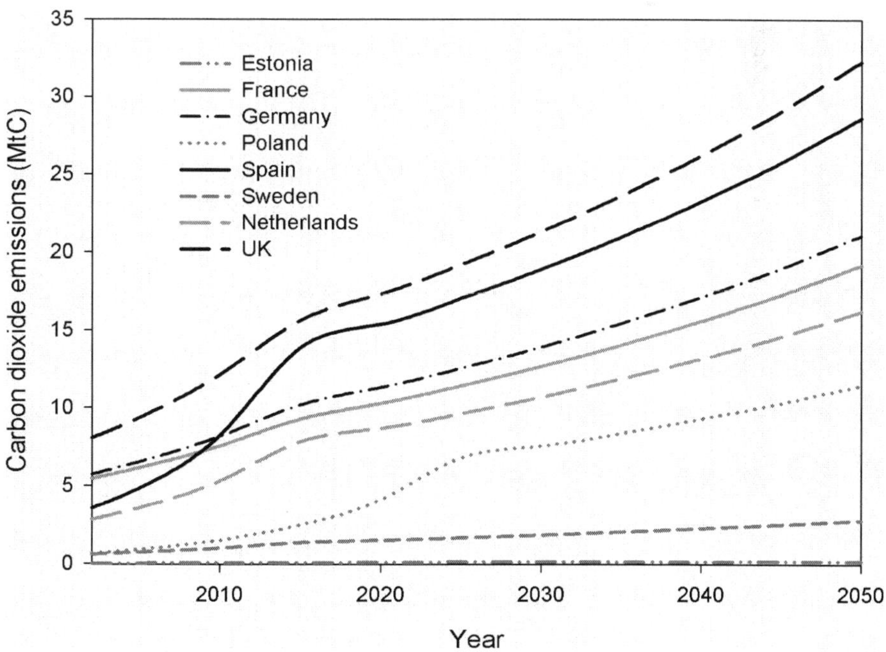

Figure 6.2 Scenario CO_2 emissions from selected EU nations.

Preliminary EU Scenario Comparative Analysis

To put the *preliminary* scenarios into the context of a contracting EU CO_2 emission budget, the scenarios presented in the previous section can be compared with the emission pathways presented in Chapter 3. As these preliminary scenarios were developed prior to Tyndall's carbon budget work, these scenarios will be compared only with the conventional Contraction & Convergence emission pathways discussed in Chapter 3.

To compare the resultant CO_2 emissions from the EU's aviation industry with the contracting EU CO_2 profile, the emissions are aggregated into those from the 'old EU' nations and those from the 'new' nations. The results are presented in Figure 6.3 and Figure 6.4.

Figure 6.3 compares the original Contraction & Convergence profiles for both the 450 ppmv and 550 ppmv cases with the aviation emission scenario for the EU15 nations, showing that as time goes by, a larger proportion of the emission allowance under this regime is consumed by aviation. By 2030, over 100 million tonnes of carbon (MtC) of the 500MtC to 725MtC range (for 450 ppmv and 550 ppmv) is emitted by EU15 flights. If these emissions were to remain constant from 2030 onwards, as the DfT suggest for the UK in their capacity constrained analysis for the UK,[7] these static EU aviation emissions would by 2050 account for some 59% of the 450 ppmv target and 29% of the

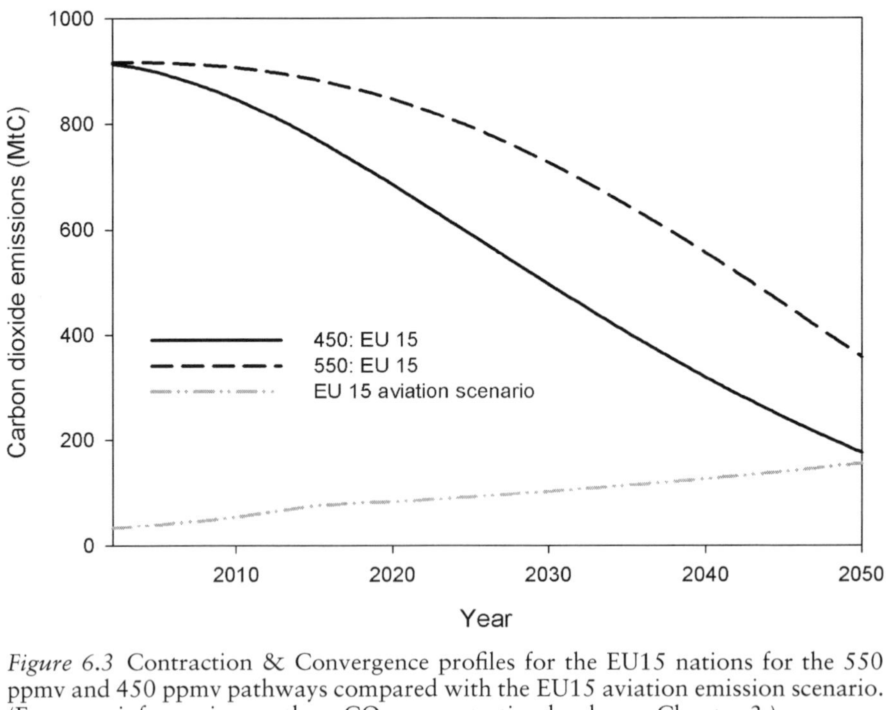

Figure 6.3 Contraction & Convergence profiles for the EU15 nations for the 550 ppmv and 450 ppmv pathways compared with the EU15 aviation emission scenario. (For more information on these CO_2 concentration levels, see Chapter 3.)

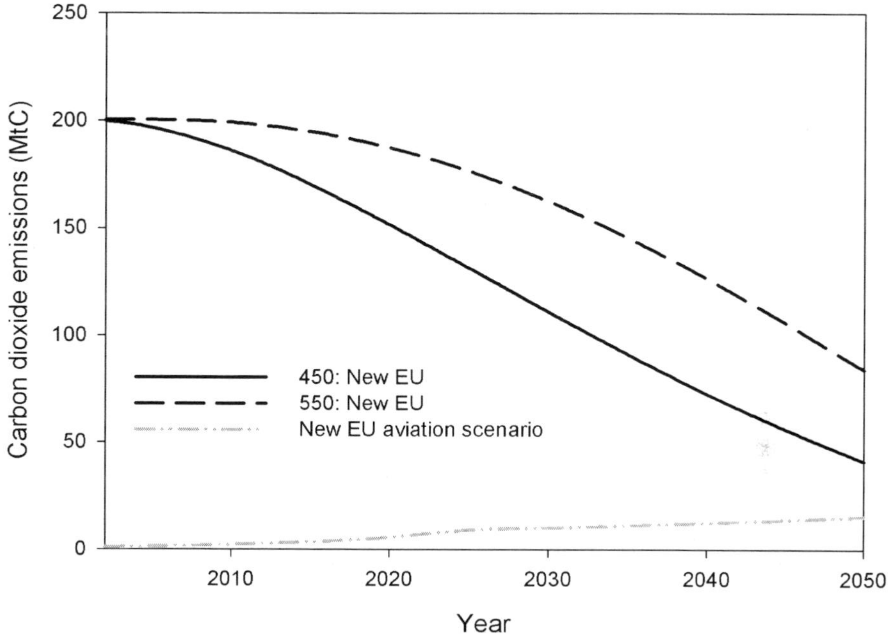

Figure 6.4 Contraction & Convergence profiles for the 'new EU' nations for the 550 ppmv and 450 ppmv pathways compared with the 'new EU' nations' aviation emission scenario.

550 ppmv target. However, if emissions continued to rise at the rate assumed for a 'mature' market, by 2050 almost all of the 200MtC of emissions for a 450 ppmv future would be emitted by aviation, and some 44% of the emissions for a 550 ppmv future. These figures are for CO_2 alone and do not include any additional emissions of NO_x or impacts on contrails or cirrus clouds.

Figure 6.4 repeats the comparison of emissions growth and Contraction & Convergence curves for the 'new' EU nations. As aviation emissions from these nations start from a much lower aviation emissions base, the assumed growth rates result in a lower proportion of the collective carbon emissions budget being consumed by aviation than for the 'old EU' nations, despite the substantial reductions in carbon emissions experienced by eastern European economies as a whole during the first half of the 1990s. For the 'new' EU nations, aviation emissions in 2030 are around 10% of the 450 ppmv total for 2050 and 6% of the 550 ppmv total for 2050. However, this needs to be seen in the context of the air-transport sector of the 'new EU' nations accounting for only 0.1% of total emissions in 2000. If emissions continue to rise beyond 2030 and up to 2050, aviation CO_2 emissions would account for just less than 50% of the quantity prescribed by a 450 ppmv profile. Figure 6.5.

Similar to Figure 3.3 in Chapter 3, Figure again shows Contraction & Convergence profiles for 450 ppmv and 550 ppmv, but in this case for the

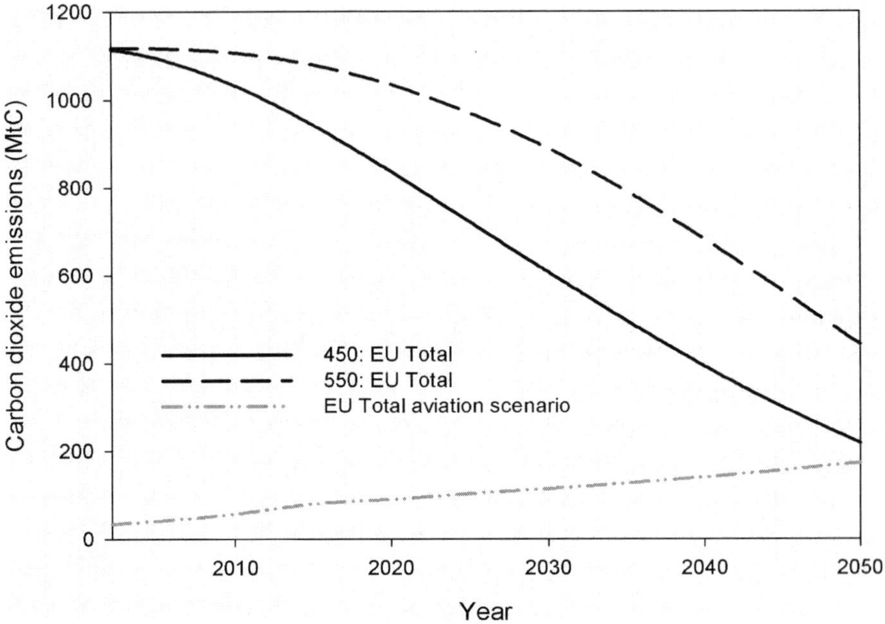

Figure 6.5 Contraction & Convergence profiles for the EU25 nations for the 550 ppmv and 450 ppmv pathways compared with the EU25 nations' aviation emission scenario.

whole of the EU. The general picture is similar to that seen for the EU15, as these nations dominate the whole European aviation scene. For 2030, emissions from EU aviation, when considering these *preliminary scenarios*, relative to 2030 Contraction & Convergence targets would account for 19% of the 450 ppmv value and 13% for 550 ppmv. If growth continued up to 2050, the industry would account for some 80% of the 450 ppmv future and 39% of the 550 ppmv future. It should again be emphasised that these values are for CO_2 emissions alone, and do not take into account any uplift 'multiplier' to represent additional radiative effects from contrails and cirrus clouds. Furthermore, the analysis above is for the *preliminary* scenarios, and presents the picture as it stood a number of years ago. The next section provides an update to this analysis, incorporating both more recent data and comprehensive input from aviation industry stakeholders.

Updated EU Aviation Scenarios

Using a recent and significant update to the work presented in the previous sections, scenarios for aviation emissions that push the boundaries of improvements in fuel efficiency and new technologies were developed. These scenarios assumed an EU successful in driving down emissions commensurate with the 450 ppmv carbon budget. Additional differences between these scenarios relate to the scenario development method, technological and operational assumptions relating to fuel efficiency, assumptions related to the use of alternative fuels, growth rates and the comparative analysis.

Scenario Method and Assumptions

Given that aviation emissions will likely be included within the EU ETS (see Chapter 5), these *updated* EU emission scenarios are considered in relation to an extended EU ETS starting in 2012. The scenarios are broken down into a number of time frames; of these, the first is the period from 2006 to the inclusion of aviation within the EU ETS, assumed to be 2012. During this period, a range of growth rates are used to reflect the range of growth rate estimates made over recent years (see Table 6.1).

In addition, a range of other factors influencing the make up of the scenarios are considered:

- the current and continued success of the low-cost air model
- access to a network of growing regional airports
- the low-cost model extending in modified form to medium- and longer-haul routes such as those between the UK and the USA
- no significant downturn in aviation between the latest 2005 data and today (2007)

Table 6.1 Selected Statistics Relating to the EU25's Aviation Industry (The years most severely affected by the events of September 11, 2001, are not included within any of the trends extracted.)

Period	*Coverage*	*Source*	*Characteristic*
2004–2005	EU25	Eurostat 2005[8]	8.5% growth in passenger numbers
2003–2004	EU25	Eurostat 2004	8.8% growth in passenger numbers
2004–2005	EU15	Eurostat 2005	4.4% growth in flights
2003–2004	EU15	Eurostat 2005	5.0% growth in flights
2004–2005	EU25	UNFCCC	6% growth in CO_2 from aviation
2003–2004	EU25	UNFCCC	6.8% growth in CO_2 from aviation
1994–2000	EU25	UNFCCC	6.3% p.a. growth in CO_2 from aviation
1990–2000	EU15	UNFCCC	5.4% p.a. growth in CO_2 from aviation
2003–2005	EU25	UNFCCC	6.4% p.a. growth in CO_2 from aviation

The data contained in Table does not provide evidence for the regularly cited 1–2% per year improvement in fuel efficiency per passenger-km. A detailed analysis of such figures would be useful, but is beyond the scope of this work and highly constrained by the inconsistency of data available across the EU25 nations.

For the years from 2006 to the end of 2011, recent and longer-term trend data significantly influence the choice of scenarios. According to the submissions to the UNFCCC, there has been a long-term trend of increasing CO_2 emissions from EU25 nations of the order of 6% per year. More recent emissions have also increased at 6% per year, once allowance is made for the period affected by the events of September 11, 2001. Reinforcing this 6% figure is EUROCONTROL's forecast of strong growth for 2007–2008.[9]

Hence, these *updated* scenarios developed here for the period from 2006 to the end of 2011, use 6% emission growth as a midrange value, with 4% per year representing the lower range and 8% per year the higher range. Due to an absence of available data in terms of passenger-km, it is difficult to relate fuel efficiency improvements to passenger-km growth rates. However, assuming no radical step changes in the short-term, the scenarios all use a 1% per year improvement in fuel efficiency across the fleet for the period 2006 to the end of 2011.

For each nation, the total domestic and international CO_2 for aviation submitted to the UNFCCC is an estimate of the CO_2 associated with all domestic flights within the EU25 and 50% of international flights to and from EU nations. This UNFCCC data gives a baseline for 2005 of 41MtC, with 7MtC from domestic flights and 34MtC from international. These figures will be different to the baseline figures used within the EU ETS, which will consider all EU-related departures and arrivals. This issue is covered in detail in Section 3.1.1. of the report *Aviation in a Low-Carbon EU*.[10] By applying the three different growth rates (4%, 6%, 8% per year) to the baseline figure of 41MtC, the emissions by 2012 range from 51MtC to 64MtC, as illustrated graphically in Figure 6.6.

In considering the *updated* aviation emission scenarios for the medium-term (2017–2030) to long-term (2031–2050), assumptions must be made in relation to both the aviation industry and the overarching EU climate policy context.

In relation to the latter, for a 450 ppmv stabilisation level, it is assumed that:

- The EU adopts a comprehensive and scientifically literate basis for its climate policy derived from a cumulative carbon budget approach.
- It has a complete account of all sectors.
- It uses a Contraction & Convergence regime[11] with a convergence date of 2050.

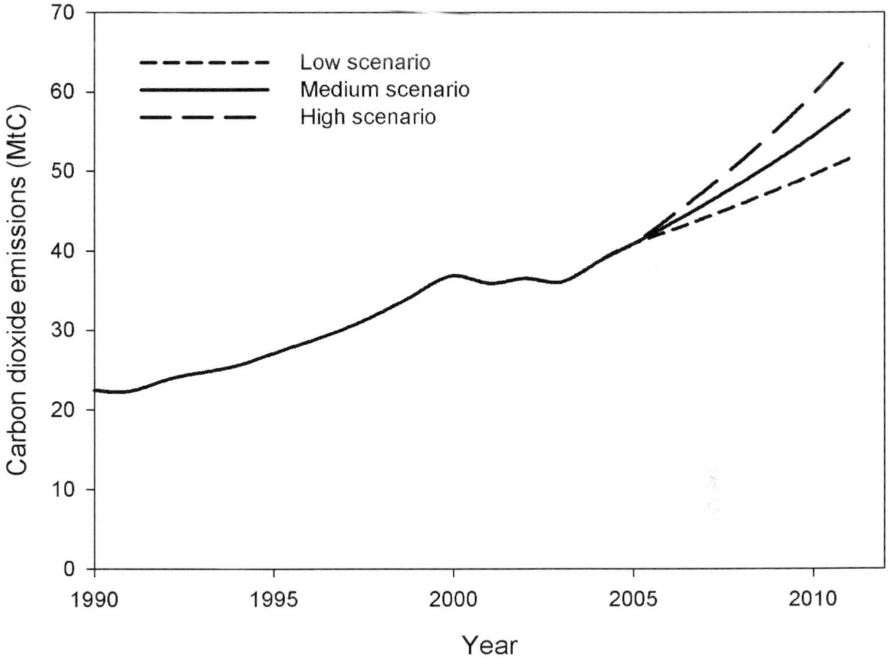

Figure 6.6 Data submitted to the UNFCCC for CO_2 emissions from EU25 nations' aviation industries from 1990 to 2005 (solid black line until 2050), and three aviation emission scenarios from 2006 to 2011.

For the aviation sector, three core scenarios commensurate with 450 ppmv are considered alongside one illustrative scenario that will fall outside of the 450 ppmv regime. Each of the scenarios are developed from 2012, with differing effects on the aviation sector.[12] Given that the core scenarios are required to be commensurate with the cumulative emissions budget for 450 ppmv, the sooner the EU responds to the climate change challenge, the less demanding will be the emissions pathway from that point onwards. To address the problem of growing emissions from the aviation sector, it is assumed that key short-term dates for EU action are 2012, the date by which aviation is assumed to be included in EU ETS, and 2017, to provide the EU with a decade to fully adopt the emission implications of its own targets.

Developing *updated* aviation emission scenarios requires the following factors to be quantified:

- The rate of growth in the near-term (i.e., 2007 to 2012);
- The rate of growth in the short-, medium- to long-term (i.e., after 2012); and

- The rate of introduction of new technologies and operational measures that may act to improve the efficiency and carbon intensity of the industry compared to the present.

Building on the *updated* scenarios to 2012, a series of medium- to long-term *updated* scenarios are developed using these factors. These scenarios have been given neutral descriptors, a method used frequently by the Tyndall Manchester researchers when developing scenarios to avoid biasing the users of any scenario set. They are called Indigo, Aqua, Violet and Emerald.

In each case, the four scenarios are divided into three time periods after 2012:

Short-term	From the start of 2012 to the end of 2016	5 years
Medium-term	From the start of 2017 to the end of 2030	14 years
Long-term	From the start of 2031 to the end of 2050	20 years

Three of the *updated* scenarios (Indigo, Aqua and Violet) are based on an assumption that the EU is committed to a meaningful 450 ppmv carbon budget, and that aviation will play its part in that process. Consequently, all these scenarios assume significant reductions in the CO_2 emitted per passenger-km flown (CO_2/passenger-km), as presented in Table 6.2.These combine to give a reduction in CO_2/passenger-km for 2012–2050 of 68.5%. The overarching context of this reduction in carbon intensity is society's explicit and genuine commitment to a 450 ppmv pathway.

It should be noted that there is considerable uncertainty in relation to the improvements that have been demonstrated historically by the aviation industry, and future fuel-burn targets, such as those set by ACARE. For example, within the IPCC's special report on aviation, it is stated that aircraft fuel efficiency improvements have been of the order of 1–2% per year, with a 70% improvement from 1950–1997.[13] This measurement is in improvements to available *seat kilometres per kg of fuel burnt* for a new aircraft in 1997 compared with one in 1950. The corresponding figure for fuel burnt per seat-km is closer to 3% per year. In the case of the ACARE target,

Table 6.2 CO_2 per Passenger-km Improvement per Period

	Short	*Medium*	*Long*
Mean annual improvement in CO_2/pkm	1.5%	2%	4%
Total improvement of the period	7% in 5 years	23% in 14 years	56% in 20 years

a 50% improvement in CO_2 per seat-km by 2020 for a new aircraft compared with one in 2000 is called for. This is equivalent to a 3.4% per year improvement in CO_2 per seat-km. In considering these and other improvements in efficiency it is important to note:

- Seat kilometres are not equivalent to passenger-km as they do not take into account changes in the load factor of the aircraft (i.e., how full the aircraft is).
- A 1% per year increase in the number of seat kilometres per kg of fuel burnt is not the same as a 1% decrease in the number of kg of fuel burnt per seat kilometre.
- Improvements to the amount of CO_2 produced per passenger-km are affected not only by improvements in fuel efficiency but also by switching to alternative fuels.
- The percentage figures in Table 6.2 refer to the whole fleet of aircraft operating in and out of the EU, and do not apply to a single new aircraft compared with one from an earlier year.

As discussed in Chapter 4, The Greener by Design study highlights a number of areas that could offer substantial improvements in terms of the fuel burn per seat-km. For example, in the short- to medium-term, air traffic management improvements could offer an 8% reduction in fuel burn, open-rotor engines of the type currently being discussed by some carriers and manufacturers could improve fuel efficiency by some 12% and the use of lighter materials such as carbon-fibre could offer upwards of a 15% improvement. Dreamliner aircraft, produced by Boeing and due to begin operations in the coming few years, could improve the fuel burnt per passenger by some 27% compared with comparative existing Boeing models, according to Boeing sources (presumably through a mixture of new materials, minor modifications to the airframe design and improved engine efficiency).

In the longer-term, laminar flow-type aircraft designs could significantly reduce fuel burn, and alternative fuels, although generally thought unlikely to be used across the fleet prior to 2030, could play a role in reducing aviation's CO_2 emissions. It is the timescale over which the gains in fuel efficiency and the incorporation of new low-carbon fuels occur that is of key importance, as this has a knock-on affect on how quickly improvements are manifested across the aircraft fleet.

In terms of these updated scenarios, technological improvements in efficiency, coupled with a variety of air traffic management and operational changes, provide the principal components of the reduction in CO_2 per passenger-km during the first two periods (2012–2017 and 2018–2030). Typical changes required to achieve this might include continued incremental jet-engine improvements, the incorporation of rear-mounted open-rotor engines particularly for shorter-haul flights and airframe modifications to improve fuel burn. It is also assumed that there will be additional load-factor increases and a series of efficiency gains across the air traffic management system through more direct routing, reduced taxiing, waiting and circling, and reduced use of the auxiliary power unit.

Fuel-switching is assumed to be a minor component within the two earlier periods, but increases significantly in the third period (2031–2050). In this long-term period, fuel efficiency improvements across the fleet are assumed to continue at around 2% per year, with an additional 2% reduction in CO_2 emissions being derived from fuel-switching to a low-carbon fuel such as biofuel, or possibly hydrogen.

The efficiency savings and low-carbon fuels assumed here reflect a situation where the aviation industry goes well beyond its achievements over the previous two decades. Such significant improvements to the technical, operational and managerial efficiency of aviation are only considered possible when driven by a concerted effort on the part of the industry (and society) to deliver them. The incorporation of low-carbon fuels in the post-2030 period will, in addition to incremental technological and operational development, require a considerable amount of innovation and fleet-wide take-up.

In terms of drivers for such change, the three scenarios reflect a society with a very strong emphasis on tackling climate change. Within such a society, low-carbon innovation would receive very significant funding and policies would be in place to regulate low-carbon behaviour and operation within companies. The difference in emphasis of this world from the current situation is central to these *updated* scenarios. Therefore it is worth reiterating that the carbon intensity improvements within the scenarios are well in excess of what has occurred within most fleets in recent times, but are in keeping with what is possible[14] if an appropriate suite of incentives were in place.

In terms of the other variables reflected in the scenarios, while the three scenarios (Indigo, Aqua and Violet) all have the same level of carbon intensity improvement, each differs in the rate of passenger growth. These factors combine to produce different emission changes between 2012 and 2050 which, in combination with the low, medium and high growth scenarios for the near-term (to 2012) set out earlier, produce a range of net CO_2 emissions from the aviation industry.

A fourth scenario (named Emerald) differs from the others in terms of both passenger–km growth and efficiency improvements. This scenario reflects only partial commitment to both curbing passenger growth rates and instigating the technological efficiency improvements described above, and is highly unlikely to be compatible with a 450 ppmv pathway.

Scenario Parameters and Emissions

Within this subsection, each scenario with its embedded assumptions is outlined.

Table 6.3 Indigo Passenger-km Growth and Carbon Intensity Improvements

Indigo	*Short*	*Medium*	*Long*
Annual passenger-km growth (pkm)	3%	1.5%	1%
Annual CO_2/pkm (op and tech)[15] improvement	1.5%	2%	4%[16]
Annual emissions change	1.5%	–0.5%	–3%

Indigo

Within this scenario aviation is very responsive to the climate change issue and the role of a meaningful EU ETS. Industry demonstrates a significant, comprehensive and early drive towards a low-carbon aviation industry within a low-carbon EU. Net aviation emission change between 2012 and 2050 equates to a *45% reduction*, though compared with 1990, represents a 24% to 55% increase. Table 6.3.

Aqua

Here aviation responds more slowly to the EU ETS scheme, compensated by slightly larger reductions by other sectors. Net aviation emission change between 2012 and 2050 equates to a *16% reduction*, though compared with 1990, it represents a 95% to 144% increase. Table 6.4.

Violet

The aviation industry's emissions continue to grow at a higher rate than in the Indigo and Aqua scenarios at the expense of the other sectors. The net aviation emission change between 2012 and 2050 equates to a *26% increase*, and compared with 1990, a 184% to 256% increase. Table 6.5.

Emerald

This additional scenario is used to illustrate a future where the current rhetoric on climate change is only partially converted into meaningful action.

Table 6.4 Aqua Passenger-km Growth and Carbon Intensity Improvements

Aqua	*Short*	*Medium*	*Long*
Annual passenger-km growth	4%	3%	2%
Annual CO_2/pkm (op and tech) improvement	1.5%	2%	4%[16]
Annual emissions change	2%	1%	–2%

Table 6.5 Violet Passenger-km Growth and Carbon Intensity Improvements

Violet	*Short*	*Medium*	*Long*
Annual passenger-km growth	5%	4%	3%
Annual CO_2/pkm (op and tech) improvement	1.5%	2%	4%
Annual emissions change	3.5%	2%	–1%

Such a future would be more attuned to cumulative emissions associated with much higher CO_2 concentrations and a failure to respond to the 2°C commitment. In this case, the net aviation emission change between 2012 and 2050 equates to a *146%* increase, and compared with 1990, a 278% to 373% increase.

Within Emerald, a modified version of the low-cost model is also assumed to extend to medium- and long-haul flights. Point-to-point aircraft (Boeing Dreamliner for example) in combination with the expansion of regional airports is assumed to provide much quicker and convenient air travel for all. Security becomes less of an obstacle to flying and big improvements in check-in improve the quality of experience for the traveller. Furthermore, flying expands to increase the number of flyers from the C, D and E social groups. Increasing globalisation stimulates more migration and consequently international travel to maintain family ties. In economic terms, world GDP growth continues and the EU exits its period of economic stagnation with its economy growing at 2.5 to 3% per year. Although it is impossible to paint an accurate picture of a business-as-usual future for aviation emissions, the Emerald scenario represents the closest to an extrapolation of current trends of all the scenarios (with the assumption that the impacts of climate change are small). Table 6.6.

Scenario Emissions

When combined with the three near-term growth scenarios (low, medium and high), the scenarios described result in a full set of nine core scenarios, with

Table 6.6 Emerald Passenger-km Growth and Carbon Intensity Improvements

Emerald	*Short*	*Medium*	*Long*
Annual passenger-km growth	6%	5%	3%
Annual CO_2/pkm (op and tech) improvement	1%	1.5%	2%[17]
Annual emissions change	5%	3.5%	1%

a further three for the Emerald set. The resulting CO_2 emission pathways for all twelve scenarios are presented in Figure 6.7.

Updated EU Scenarios—Comparative Analysis

In terms of a comparison with total EU emissions, the *medium* aviation scenarios are plotted against the 450 ppmv carbon pathways in Figure 6.8.

Figure 6.8 illustrates that, unless very low growth rates and substantial improvements to carbon efficiency are achieved, aviation emissions will likely exceed the 450 ppmv 'low' pathway by the late 2040s (i.e., the Violet scenario exceeds the 450 ppmv 'low' curve, whilst Indigo is within the 2050 budget). For the 450 ppmv 'high' pathway, the emissions from aviation account for at best 10% and at worst 29% of the total budget for all sectors and all emissions. If, on the other hand, emissions continue to grow at a rate closer to recent trends, as in the Emerald scenario, then the compensation required by other sectors to remain within a 450 ppmv target will either be impossible (i.e., exceed the 450 ppmv 'low' pathway), or extremely challenging (i.e., represent over 50% of the 'high' 450 ppmv pathway). However, given the 450 ppmv CO_2 pathway is associated with a 70% chance of exceeding the 2°C temperature threshold, remaining within the 'low' 450 ppmv pathway is highly desirable from the perspective of the climate change objectives the EU has set for itself (i.e., not exceeding the 2°C threshold).

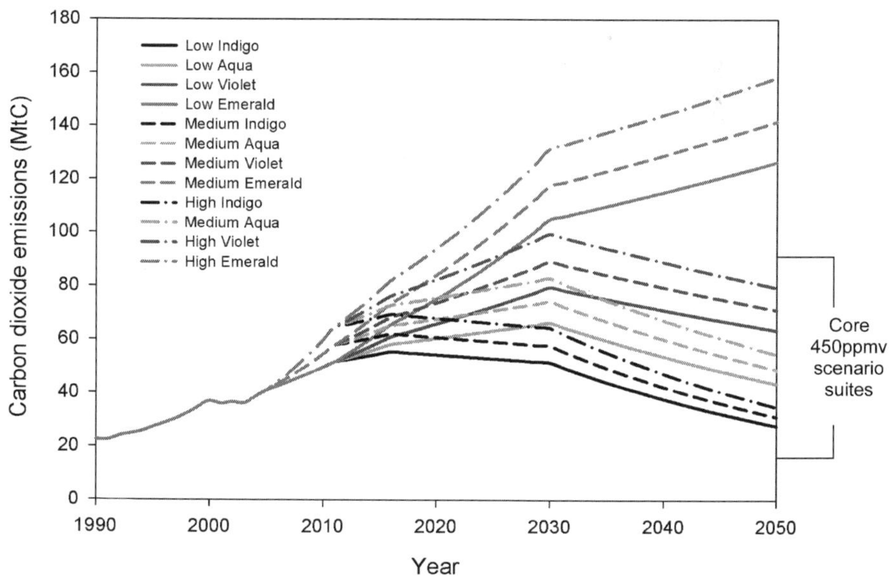

Figure 6.7 CO_2 emissions from the nine core scenarios (Indigo, Aqua and Violet for pre-2012 near-term growth scenarios low-1, medium-2 and high-3) and the three illustrative higher growth scenarios (Emerald for pre-2012 near-term growth scenarios low-1, medium-2 and high-3).

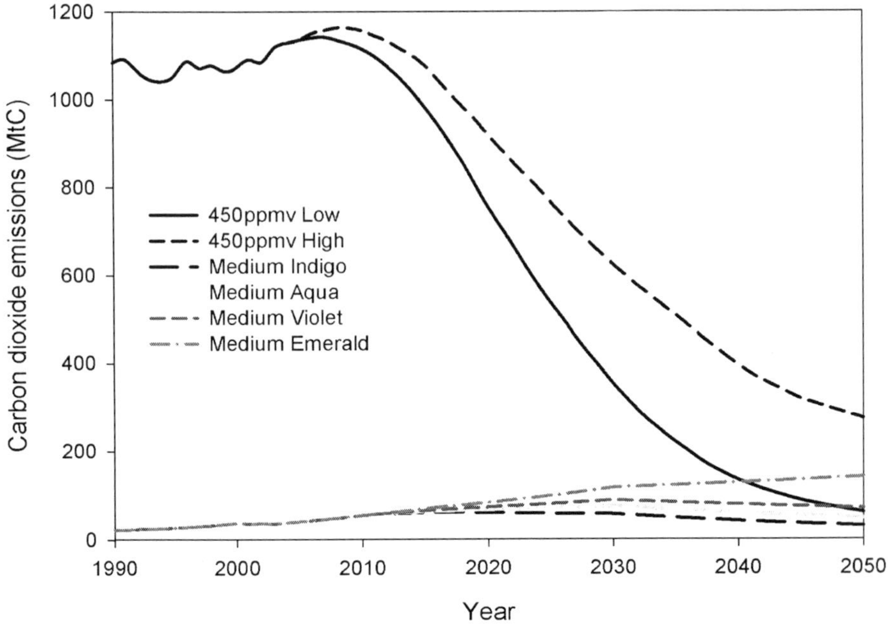

Figure 6.8 CO_2 emission budgets for 450 ppmv compared with the mid-growth aviation emissions scenarios based on the UNFCCC data to account for 50% of international flights and all domestic and intra-EU flights.

It is clear from the analysis that all scenarios of the aviation industry, even within a world striving to achieve a 450 ppmv future (Indigo, Aqua and Violet), reflect a large increase in the industry's CO_2 levels in 2050 compared with 1990. Without a sea change in attitude towards aviation, the industry's emissions, under all the scenarios, have major impacts on the already challenging carbon reductions demanded of other sectors. Potentially, aviation growth and a commitment to 2°C could require all other sectors to be completely decarbonised by 2050. This is in sharp contrast to the other sectors of the economy, where 75% to 90% reductions from 1990 levels have been required.

AVIATION EMISSION SCENARIOS FOR THE UK

It is often argued by the aviation industry that it is inappropriate to compare national carbon budgets with aviation's CO_2 emissions due to aviation's international nature. This argument was voiced loudly when documentation of the UK's aviation emissions compared with a UK carbon budget was published in 2005,[18] with calls for comparisons only at a global level. Given the international nature of aviation, this may appear to be a valid criticism. However, similar arguments can also be made for shipping and road transport across continental Europe. In a globalised world, many emissions have international drivers, for example, the city of London provides a global service, pharmaceuticals, chemicals, steel and aluminium all have international markets. If the international dimension of aviation is to be a major factor in apportioning emissions, it should only be done so after careful consideration of the international dimension of much of the UK's and EU's economy. Moreover, the reality of aviation is that very few of the world's population actually fly, with the vast majority of flying occurring within OECD nations.

Arguments in favour of presenting aviation's emissions in the context of either the EU or UK carbon budgets are grounded on the basis that policy-makers require an understanding of future climate change drivers and mitigation opportunities. If individuals in industrialising nations were to begin flying as often as UK citizens, and the world made a concerted effort to cut emissions over time globally, the emissions from aviation compared with all of the other sectors would be even more significant than they are already. Furthermore, the rate at which aviation emissions continue to grow in industrialised nations indicates that the market is some years from maturity.

By being able to compare CO_2 emissions from UK-related aviation with the UK's total CO_2 emissions, and producing scenarios of possible future CO_2 levels within a contracting carbon budget, the scale of the challenge faced not only in the UK, but in many industrialised nations, can be illustrated. Understanding the significance of one rapidly growing sector against the backdrop of sectors whose emissions are, in general, and with the exception of international shipping, currently broadly static, or at low levels of growth (e.g., road transport), provides clear lessons for nations whose aviation industries are only just beginning to play a significant role in transportation.

To compare the UK's CO_2 emissions from aviation with the UK's total carbon budget, two separate analyses are presented here. Firstly, aviation scenarios developed in 2005 for comparison with the UK Government's own scenarios (presented in Chapter 5) are compared with the original Contraction & Convergence profile consistent with the UK Government's 60% carbon-reduction target (Figure 3.5). Secondly, scenarios drawn out of the EU analysis presented in Section 6.2, but for the UK alone, are compared with the more realistic carbon profile consistent with the 2°C target (Figure 3.6).

Preliminary UK Aviation Scenarios

Three scenarios for the UK's aviation industry are presented here, providing a range of outputs for the period 2004 to 2050. The first scenario is based on the same methodology outlined in Section 6.2, with the other two scenarios developed to provide a contrast. Scenarios assumptions are:

Scenario 1—Government in the Know!

Passenger-km growth:
2004–2015—continuation of pre-2001 trend of 6.4%[19] per year
2015–2050—reduction to 3.3%

Fuel efficiency improvements:
2004–2050—1.2% per year

Scenario 2—Market Soon Matures

Passenger-km growth:
2004–2010—continuation of pre-2001 trend of 6.4% per year
2010–2050—reduction to 3% representing market maturity

Fuel efficiency improvements:
2004–2050—1.2% per year

Scenario 3—Europe Rules

Passenger-km growth:
2004–2010—continuation of the current trend of 7% per year
2010–2030—reduction to 4% per year in line with European forecasts
2030–2050—reduction to 3% per year representing market maturity

Fuel efficiency improvements:
2004–2010—1.7% per year in line with BA target
2010–2050—1.2% per year

The first scenario is essentially in line with UK Government forecasts, but with an update to take account of the growth both prior to the events of September 11, 2001, and the resurgence witnessed in the last couple of years. The second scenario is more optimistic in terms of emissions, suggesting a significant drop in the growth rate by 2010, indicating a major downturn in the industry.

The final scenario uses highly optimistic fuel efficiency improvements until 2010, suggesting that the whole fleet operating in and out of the UK has improved fuel efficiency at BA's target rates. In terms of growth, a continuation of current levels is again curtailed to European projected growth rates until 2030. Following this period, the industry is assumed to be mature.

The impact of these growth scenarios on UK aviation CO_2 emissions is illustrated in Figure 6.9. The difference between the early years of Scenario 3 and the other two scenarios is due to this scenario being produced after the release of the latest energy consumption data for 2004 from the Digest of UK Energy Statistics,[20] which indicates aircraft bunker fuel consumption as having risen by 10% between 2003 and 2004. The range of emissions exhibited by the scenarios in 2030 is 17.3MtC to 22.5MtC, and at 2050 is

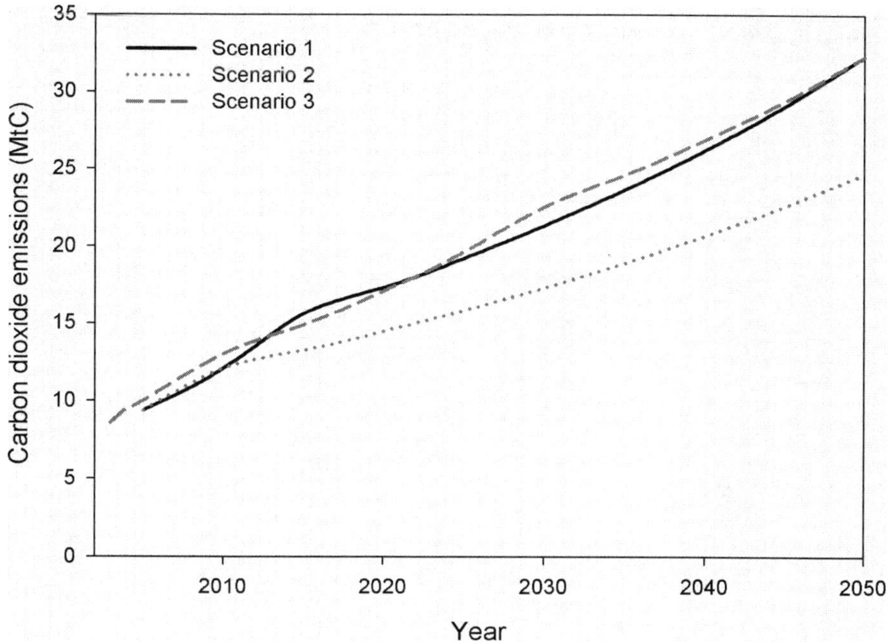

Figure 6.9 Three Tyndall carbon emissions scenarios for UK aviation.

between 24.7MtC to 32.3MtC. Scenarios 1 and 3 are therefore close to the DfT's 'worst' case scenario presented in the previous chapter, with Scenario 2 bridging a gap between the 'worst' case and the 'central' case. In none of these scenarios do the emissions level off. Without deliberate policy decisions for substantially curbing the rate of air traffic and passenger growth, there is no reason to assume that the industry's emissions will stop growing within the time frame of this analysis (i.e., to 2050).

Preliminary UK Aviation Scenarios—Comparative Analysis

Employing the same approach taken in Chapter 5 when comparing the DfT's forecasts with the original Contraction & Convergence profiles, these UK aviation scenarios are compared with the 450 ppmv and 550 ppmv CO_2 pathways available at the time of the publication of the Aviation White Paper. The scenarios demonstrate that unless there is an unforeseeable step change in technology or operations, or a significant reduction in growth by 2020, the range of carbon emissions generated by aircraft is to be between 14.5MtC and 17.3MtC. This compares with the range of 13.4MtC and 16.5MtC in the DfT's forecasts in Figure 6.10.

In relation to the 550 ppmv curve, aircraft contribute between 10% and 12% of the total carbon budget in 2020. Relative to the 450 ppmv profile, by 2020 aviation consumes between 13% and 15% of the total carbon

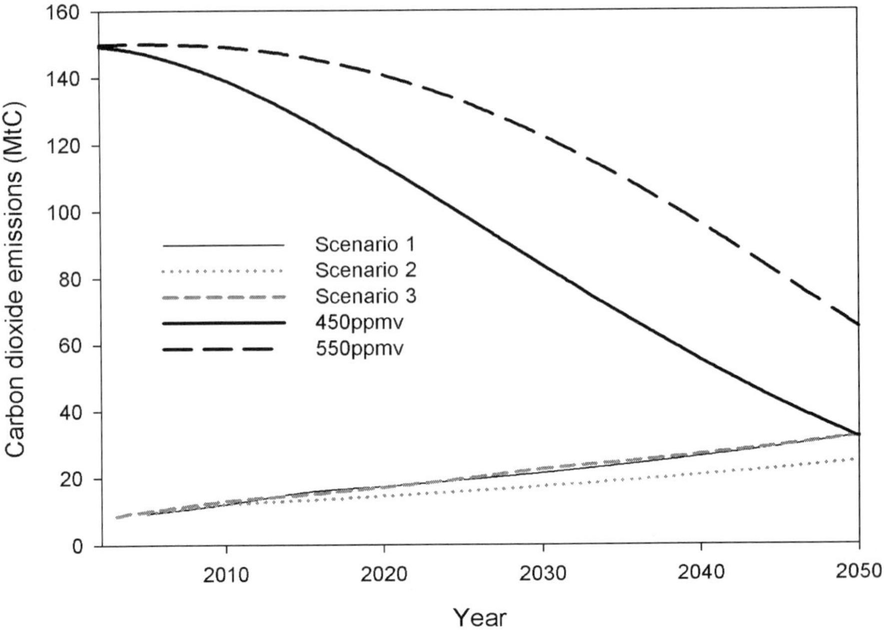

Figure 6.10 UK 450 ppmv and 550 ppmv Contraction & Convergence profiles compared with three UK aviation CO_2 emission scenarios.

budget. By 2030, between 14% and 18% of the budget is consumed under a 550 ppmv profile, and under a 450 ppmv target, between 21% and 30% of the total carbon budget is consumed. By 2050, only Scenario 2 stays within the 450 ppmv pathway, but even this scenario consumes 77% of the total carbon budget for 450 ppmv. Under the 550 ppmv profile, the range taken up by the aviation scenarios is between 38% and 50% of the total. Note that these profiles are for CO_2 emissions alone.

The results presented here point to a conflict between the UK Government's aviation and energy policies. Furthermore, as discussed earlier, if the UK Government is to develop policy consistent with its own 2°C target, the 550 ppmv CO_2 profile can no longer be considered appropriate.

Updated UK Aviation Scenarios—Comparative Assessment

The final section of analysis within this chapter focuses on the most recent aviation scenarios developed for the UK, in comparison with the 450 ppmv CO_2 profile consistent with the UK's own 2°C target, as presented in Figure 3.6. In this case, the UK scenarios are developed with the same assumptions as for the EU. These scenarios therefore represent a highly optimistic view of the aviation industry in relation to CO_2 emissions, apart from for the Emerald scenario, as discussed previously.

To reflect on how a radically more efficient aviation industry within the UK compares with the UK's overall carbon budget, the emissions presented in Figure 6.11 are compared with the UK's *high* and *low* 450 ppmv carbon profiles first illustrated in Figure 3.6. The results are illustrated in Figure 6.12. The significance of the proportion of emissions generated by the UK's aviation industry in comparison with a contracting carbon pathway commensurate with the UK Government's own 2°C temperature threshold target is immediately striking, even for these optimistic aviation emission scenarios. The 450 ppmv *low* profile is exceeded by all scenarios by the mid-2040s, and only the Aqua and Indigo scenarios, with their substantially lower growth rates, remain within the 450 ppmv mid-budget by 2050. The Emerald scenario is completely incompatible with the entire range of 450 ppmv profiles before 2050, and the Violet scenario remains only within the 450 ppmv *high* profile. These results have serious implications for policy-makers genuinely committed to the 2°C target. Only the lower growth rate scenarios offer an opportunity to remain within the 450 ppmv budget, and even then only if other sectors are able and willing to compensate significantly for aviation's emissions. This is assuming emission reductions cannot be made through some alternative mechanism, such as the Clean Development Mechanism. Implications of such scenarios for other sectors in the economy are discussed in Chapter 7.

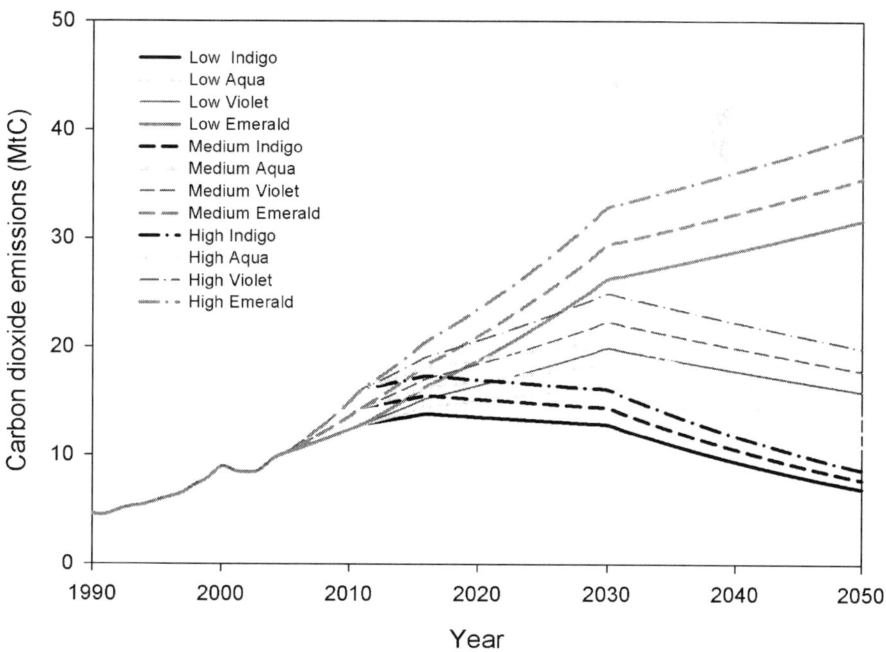

Figure 6.11 UK aviation CO_2 emission scenarios based on the assumptions presented in this section.

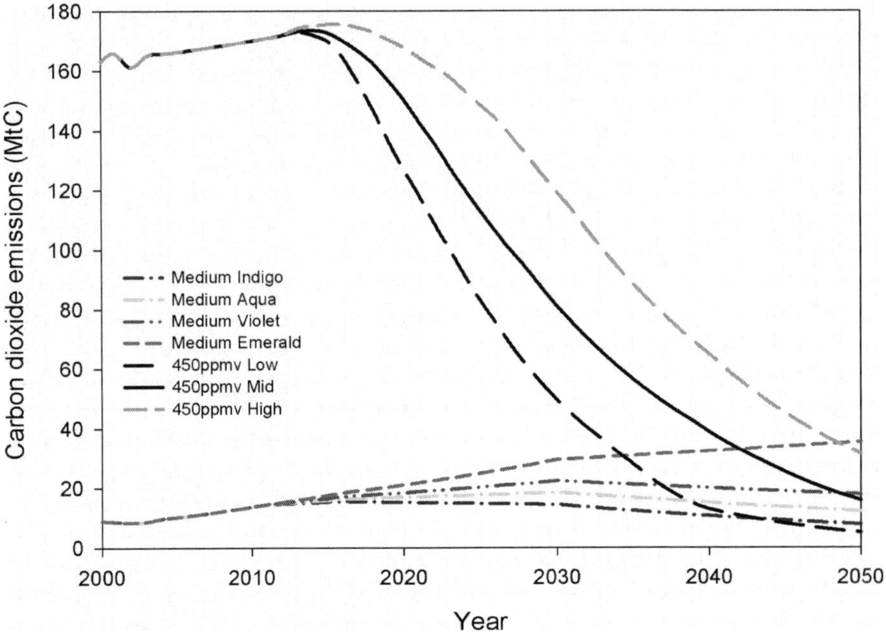

Figure 6.12 Contracting CO_2 pathways for the UK compared with the updated UK aviation CO_2 scenarios for the medium level growth rate until 2012, followed by assumptions presented previously.

SUMMARY

This chapter draws together aviation emissions and climate science, technology and policy to illustrate the scale of the challenge faced in addressing the growing CO_2 emissions from aviation. By considering possible technological and operational options, together with a range of growth rates for passenger-km, a scenario approach allows a comparative analysis of aviation's CO_2 emissions with CO_2 profiles in keeping with both the EU and UK's 2°C temperature threshold target. One of the issues discussed in Chapter 3 is the challenge of early mitigation in order to prevent dramatic and potentially economically disastrous year-on-year emission cuts as the century progresses. Whichever way the cake is cut, there is no avoiding a requirement for deep cuts in emissions by 2050, as illustrated in Figure 3.6. The aviation sector is not immune to these concerns, and has come to acknowledge this fact only very recently. This recognition is crucial given that within the EU, only one other sector's emissions appear to be growing at a rate similar to that from the aviation sector—namely, shipping. Moderate technological and operational modifications within the aviation industry will not be adequate to curb emission growth given the current high and increasing rates of demand in terms of passenger-km. Members of the aviation industry attending a recent Tyndall workshop voiced this concern, and suggested that the industry may need to be challenged to rapidly accelerate innovation and technology roll-out, if it is to stabilise emissions in the coming decades, let alone make the kinds of cuts required of other industrialised sectors. These new voices within the industry are very welcome given the scale of the challenge presented within this chapter.

Clearly, if aviation emission growth is to be addressed, in addition to pushing the bounds of technology and operational practices, investment in expanding aviation infrastructure must be questioned. There are a number of key policy areas which are of relevance to EU nations as climate and transport policies are developed and modified over the coming decade:

- The decision as to whether to include international emissions in domestic carbon-reduction targets
- The decision as to whether or not national or EU policies should be bounded by science-based cumulative emission pathways, or politically expedient 2050 targets
- The decision as to whether to tighten constraints on aviation via taxes, charges and/or limits on airport infrastructure supply
- Whether to apply strong carbon-reduction targets to other sectors, both in the UK and the wider EU, independently of EU ETS[21]
- The extent and pace at which the aviation industry is brought within the EU's emissions trading scheme
- Whether or not international emissions are allocated to nations in a post-Kyoto regime
- The possibility of trading international aviation emissions within a global emissions trading scheme, and whether such trading actually reduces net emissions in the absence of international caps

Many within the aviation industry believe incorporating aviation into the EU ETS will be the 'silver bullet' for aviation's impact on climate change. The results presented earlier for a 450 ppmv CO_2-constrained EU suggest that, if this is to be the case, the carbon price must become considerably higher than current levels. New drivers to accelerate innovation considerably will not be adequate to remain within the budget without some moderation in growth rate. To keep within the 450 ppmv CO_2 budget through the use of the EU ETS, substantial rises in air ticket prices will be unavoidable.

Importantly, the reader should be reminded that within this comparative assessment, the additional climate impacts of aviation in the form of warming caused by contrails, cirrus clouds and other emissions are not considered. These additional effects have historically impacted on the climate around one to two times more than the CO_2 alone. This does not, however, mean that this will be their relative impact in the future.

The fact that aviation's CO_2 emissions already (in 2007) account for around 6–7% of the UK's emissions and 3–4% of the EU's total CO_2

emissions, coupled with very high rates of growth, directs the climate mitigation spotlight at international aviation—although the likelihood is that international shipping activity will be increasingly drawn into such debates. Some within the industry voice opinions that such a focus is unreasonable and unfair. One misunderstanding currently appears to revolve around opportunities for change in other economic sectors. It has been suggested that a focus on aviation has ignored the high proportion of emissions from, for example, the car sector. The next chapter will illustrate why this is not the case, and further stress the importance in recognising the constraints within which the aviation industry must act to mitigate its emissions. The lengths to which other sectors and other nations can feasibly mitigate their CO_2 emissions will essentially determine the 'CO_2 space' remaining for the aviation sector. Chapter 7 will now present a brief review of mitigation opportunities outside of the aviation sector.

7 Aviation in the Wider Energy Context

INTRODUCTION

The six previous chapters have described a growing aviation industry within the context of a world genuinely committed to tackling climate change. The potential implications of a rapidly growing aviation industry for other sectors of the economy and society have been alluded to on several occasions. Within this chapter, a more integrated perspective of the UK's energy system is presented to illustrate mitigation opportunities in other sectors, and highlight why the aviation sector, in particular, has warranted so much interest since climate change rose up the political agenda.

THE TYNDALL SCENARIO METHOD

The Tyndall Centre report, *Decarbonising the UK*,[1] summarised research exploring a range of technical, managerial and behavioural options for bringing about a true 60% reduction in the UK's CO_2 emissions by 2050. This report contained a new set of integrated energy scenarios, articulating alternative carbon-constrained futures. These scenarios were unique as the first to fully include both international aviation and shipping emissions, thus providing UK policymakers with the first economy-wide illustrations of the scale of climate change challenge faced in meeting the UK's 60% CO_2 2050 target. Tyndall began its Phase 2 research in 2005, at which point the original scenarios method was subsequently developed to produce a further set of scenarios, in keeping with the 450 ppmv emission pathway presented in Chapter 3, Figure 3.6. Tyndall's Integrated Scenarios method is currently being used to build scenarios for the EU25 and China. Although continually updated and improved, the original methodology developed remains the basis for Tyndall's exploration of energy systems.

This chapter describes the Tyndall Integrated Scenarios methodology and Tyndall's Decarbonisation Scenarios produced in 2005, which adhere to the UK's 60% CO_2 reduction target. In addition, results from the more up-to-date set of scenarios commensurate with the cumulative 450 ppmv pathway are presented. This scenario work was carried out by a similar team of researchers to the team conducting Tyndall's aviation work, and therefore provides a concrete grounding for placing aviation in a wider context. The chapter will begin by considering Tyndall's work in the context of other energy research. For more information on the Tyndall scenario methodology and results, see various papers and results posted on www.tyndall.ac.uk/publications or www.tyndall.manchester.ac.uk.[2]

THE ENERGY POLICY CONTEXT

Throughout the past decade, a range of energy scenarios have informed UK policymakers; these have, without exception, provided only a partial account of the UK's energy system as they have all failed to include the two fastest growing sectors of the UK's economy—international aviation and shipping. For example, energy scenarios developed by the Performance and Innovation Unit (PIU) in its Energy Review, and used as an input to the UK's Energy White Paper,[3] considered only emissions from domestic aviation and inland and coastal shipping when analysing the UK's aviation and shipping industries. Furthermore, the method dominating the generation of energy scenarios within the UK has, and continues to be, based upon a twin-axis framework[4] which, it can be argued, has tended to constrain opportunities for generating more creative solutions to the UK's energy problems.[5] Consequently, whilst the UK energy landscape appeared to be well populated with energy scenarios, the Tyndall Integrated Scenario method, combined with the inclusion of all major emission sources, provided an opportunity to develop the first comprehensive energy scenarios consistent with the UK Government's 2°C temperature threshold.

The UK Government's decision to disregard emissions from international aviation and shipping is in line with international agreements and frameworks, such as the Kyoto Protocol. However, although within the UK these sectors are not currently the largest in terms of their overall energy consumption or CO_2 emissions, they are the two fastest growing emission sources within the economy and therefore cannot be ignored given that the ultimate objective of climate change policy is to stabilise global atmospheric CO_2 concentrations. Furthermore, as demonstrated in earlier chapters, if the aviation sector were to continue to grow at rates similar to its historical trend, then without a step change in technology, it would become one of the most important emission sources by 2050. Similarly, in a world with increasing international trade, most of it transported by ship, emissions from international shipping will likely represent a significant proportion of any constrained international carbon budget. Given the potential scale of emissions from these sectors, international negotiations must include emissions from international aviation and shipping if they are to genuinely address atmospheric CO_2 concentrations. Any analysis which excludes these emissions substantially distorts the policy message and significantly underestimates the changes required to achieve a sufficient level of decarbonisation.

SCENARIO METHOD

Scenarios are a means by which potential futures may be explored, and in so doing provide a framework intended to help stakeholders to think about the future and the processes that may shape it. The strength of the scenario process lies in the limitless variety of driving forces that may be considered, including technology developments, societal changes, policy implementation and environmental change. Moreover, the process not only includes quantifiable parameters, but facilitates a blend of qualitative and quantitative information to more fully inform future thinking. Ultimately, scenarios are not predictions or forecasts, but instead explore the possibility space through the articulation of a set of 'what-ifs'. As such, scenarios may be considered 'learning machines'; increasing the scope to more fully understand future diversity and opportunities.[6]

The Tyndall scenario approach explores mitigation options for an energy system, of which aviation is a part, by breaking the energy system down into sixteen separate demand sectors. Complementing the demand-side, is an analysis of both high- and low-carbon state-of-the-art supply options, along with efficiency and technology changes over time to reflect technological advances. Quantifying energy consumption and emissions generated by a particular demand and supply choice is possible through the development of a 'scenario generator'. This tool allows the Tyndall Manchester team to build diverse ranges of 'what-if?' scenarios that not only contain disaggregated demand data, but provide the corresponding primary energy makeup and subsequent carbon emissions. The tool can, for example, be used to explore alternative energy systems aimed at specific emissions targets in a chosen year, such as the UK Government's 60% carbon-reduction target for 2050.

To date, no other UK energy scenarios explicitly consider the transition from the present-day energy system to one that is substantially decarbonised; this is a further and important motivation for the Tyndall work. In line with the backcasting approach to scenario building proposed and developed by amongst others, Amory Lovins, John Robinson and Kevin Anderson,[7] *pathways* to alternative endpoint futures, all of which achieve a particular reduction in CO_2, can be articulated. This is in contrast to prospective scenarios that look forward and outline futures based on current trends, or extend forward a number of key drivers, usually based on 'marginal' relationships between one another. The methodology employed for developing the Tyndall scenarios defines an emission 'endpoint', builds an energy system consistent with the endpoint, and explores the pathway towards that endpoint both quantitatively and qualitatively. Figure 7.1 presents a diagram of the steps involved. Again, more information on the scenario method and full outputs can be found on the Tyndall websites.[8] Note that a scenario is not only the endpoint, but also comprises its associated pathway.

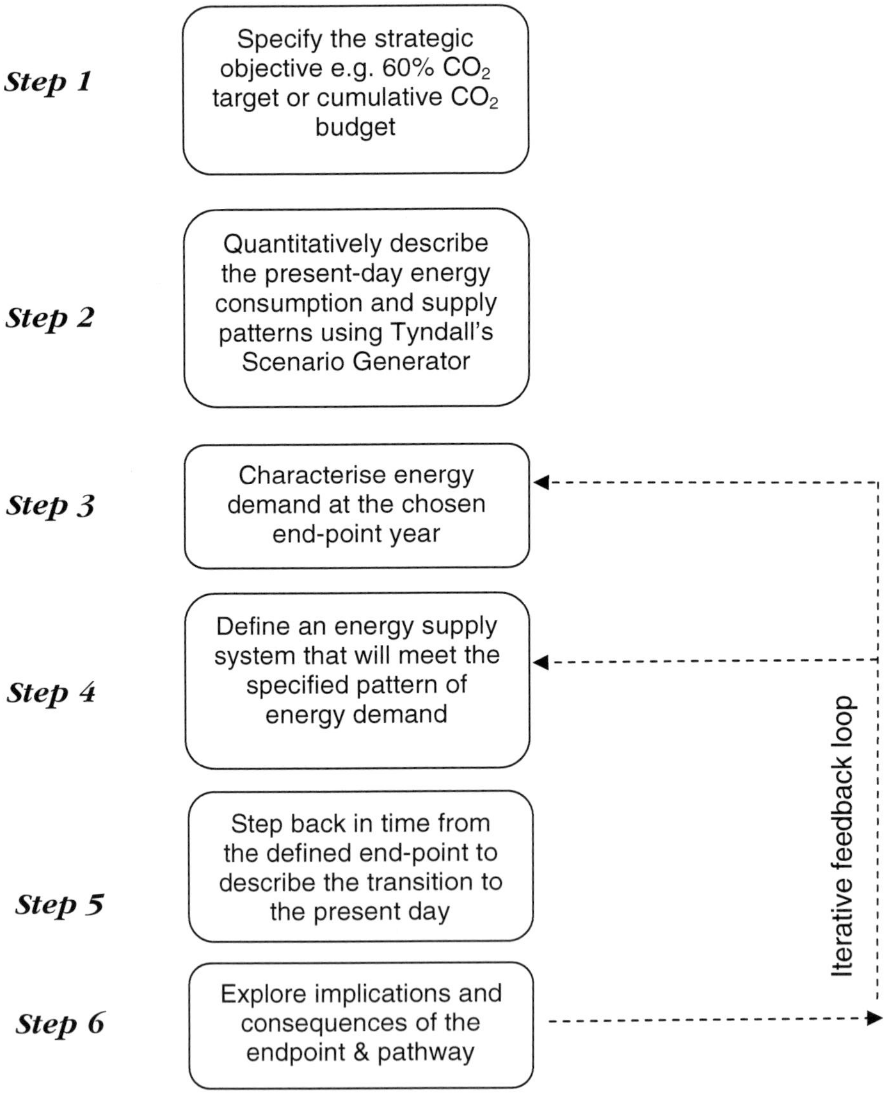

Figure 7.1 Backcasting methodology employed to generate Tyndall's Integrated energy scenarios.

TYNDALL'S 60% ENERGY SCENARIOS

Tyndall's first set of Integrated Scenarios were designed to meet an endpoint constrained by a 60% CO_2 emission reduction by 2050, in line with the targets enshrined in the UK's 2003 Energy White Paper.[9] Using the 'scenario generator', five energy endpoints achieving the 60% carbon reduction were developed. Each endpoint generates 65MtC in 2050, as opposed to an estimated 165MtC in 2002 (the baseline year for the study). The endpoints contain a detailed picture of energy supply and demand in the UK for 2050. For individual demand sectors, the energy supply system was matched to the pattern of consumption envisaged on the basis of linking energy from different fuel sources to the most appropriate end use.

To inform the range of scenarios to be developed, various levels of energy consumption in 2050 were considered by taking into account consumption levels in other UK scenario sets and historical trends in energy and economic data. The levels chosen were: 90 million tonnes of oil equivalent (Mtoe), 130Mtoe, 200Mtoe and 330Mtoe, with the current UK consumption in the region of 170Mtoe. The 90Mtoe lower limit was considered challenging but plausible from a demand reduction perspective, with the upper limit of 330Mtoe considered to be challenging from a technical and also social and political perspective. Overall, the range chosen gave rise to some scenarios requiring significant reductions in energy consumption, with others requiring an extensive low-carbon supply, and yet other scenarios with significant elements of both demand and supply changes. All of the scenarios require significant improvements in energy intensity—the energy use per unit of activity, be it passenger-km travelled or economic activity.

The scenarios were built around qualitative drivers informing quantitative energy demand and supply futures. In tandem with the quantification, the scenarios incorporated stakeholder expertise on the possible technological, behavioural and energy efficiency improvements across the household, industry, service and transport sectors as well as innovation and development spanning the energy supply system. Varying rates of economic and activity growth across the sectors gleaned from trend data and expert stakeholder judgement, coupled with the efficiency improvements assumed on the demand-side served to generate a final energy-consumption figure for each sector in the endpoint year. The pattern of energy demand was then matched to appropriate energy supply for each sector. Once both the demand- and supply-sides were quantified, the CO_2 emissions were calculated.

The scenarios subsequently underwent a process of cross-checking and confirmation through stakeholder engagement, to ensure their validity, credibility and usefulness. Furthermore, stakeholder input was used to develop pathways to the 2050 endpoints using the backcasting approach. This process resulted in identifying critical elements in relation to policy, infrastructure, and technological developments between 2002 and 2050.

The bottom-up process developed for generating these first Tyndall Integrated Scenarios resulted in a suite of five scenarios that do not lend themselves to simple characterisation. Consequently, to encourage the users of the scenarios to interpret them within a more inclusive context, they were allocated neutral descriptors. Within this report the five scenarios are referred to as Red, Blue, Turquoise, Purple and Pink, with Orange representing the present day.

Scenario Descriptions

Brief descriptions of the scenarios, derived from the output of the backcasting workshop and the project team's own analysis,[10] are set out below. Table 7.1 summarises the pertinent features of the scenarios.

The Red Scenario

The *Red Scenario* has a high level of economic growth, yet low energy demand. The UK remains primarily a service economy, with the commercial sector contributing approximately three quarters of GDP. There has been a gradual expansion of manufacturing and the public sector has declined in importance. The 60% target is met through significant energy-demand reduction, high levels of innovation to improve energy efficiency and moderate amounts of low-carbon energy supply.

Demand-Side Characteristics Demand reduction plays a much greater role in achieving the desired target than supply-side developments. For example, a high rate of technological innovation to improve the energy efficiency of appliances and buildings and a reduction in the growth rate in aviation contribute to a significant reduction in energy consumption.

Supply-Side Characteristics The 50% reduction in energy demand in 2050 compared with levels in 2002 ensures that supply-side infrastructure is kept to a minimum, despite the carbon constraint. Relatively low uptake of new supply technologies, including carbon capture and storage and hydrogen, in combination with the high demand reduction, have uncoupled the relationship between economic growth and CO_2 emissions.

Aviation Industry Whilst passenger-km travelled by plane have doubled by 2050 compared with 2002, annual growth in passenger-km in aviation has reduced to 1.4%. Changes include a reduction in real terms in business travel as a consequence of innovations in virtual technology, a reduction in short-haul flights through high-speed rail substitution and the success of market-based instruments in increasing the cost of flying. Despite aircraft using broadly similar technology, fuel efficiency has been radically improved through the implementation of a new air traffic control system for Europe, the use of lighter aircraft materials, and a quicker fleet renewal. Demand management within the aviation industry allows conventional kerosene to continue to be used worldwide. Table 7.2 summarises aviation's characteristics within this scenario. The relatively low energy consumption of the aviation industry within this scenario produces less CO_2 in 2050 compared with 2002, contributing 12% of total UK CO_2 emissions for 2050.

It is important to note that such a low rate of growth on average between 2002 and 2050 would, if current rates of growth continue for a short period of time, result in an annual reduction in passenger-km in real terms in the later years.

The Blue Scenario

The *Blue Scenario* exhibits both modest economic growth and energy demand, with the commercial sector's contribution to national wealth being almost matched by the expansion of the public sector. The lower annual economic growth is an important factor in reducing future energy consumption. However, as energy consumption is higher than in the *Red Scenario*, low-carbon supply technologies, and a more distributed low-carbon supply infrastructure is necessary to meet the 60% target.

Demand-Side Characteristics As energy demand has reduced by a quarter compared with 2002 in an economy over twice the current size, there has been an improvement in the historical trend, with regard to the energy intensity of goods and services. Politically, a strong central government establishes energy and carbon targets and policy goals, and then instructs appropriate tiers of local and regional government and accountable bodies to develop the means for implementation. This is also a highly mobile society with growth in both private and public transport.

Supply-Side Characteristics Climate change has been an overarching policy issue which has driven policy in other areas, particularly transport. Hydrogen has been promoted as a transport fuel within niche markets early on in the century. Meanwhile the low cost of coal encourages the construction of gasification with carbon capture and storage plants for hydrogen production, and an infrastructure for liquid fuel purchased at 'Hydro-stations' is in place by 2020. Hydrogen fuels take a 75% share of the road-transport fuel supply market by 2050.

Aviation Industry The aviation industry has continued to grow, but at rates around half that experienced in 2002. Improvements in fuel efficiency continue at around 1% per year. Despite the development of alternative fuels for other modes of transport, aviation remains a kerosene-consuming sector, and as a result it has become the largest carbon-emitting sector within the UK's economy, as demonstrated in Figure 7.4. Table 7.3 summarises the key aviation parameters.

Table 7.1 Summary of Tyndall's 60% Scenarios

	Red	*Blue*	*Turquoise*	*Purple*	*Pink*
UK GDP increase (per year)	3.3%	1.6 %	2.6%	3.9%	3.9%
Dominant economic sectors	Commercial	Commercial; public admin; non-intensive industry	Commercial; construction; public admin	Commercial; non-intensive industry	Commercial; non-intensive industry
Energy consumption (Mtoe)	90	130	200	330	330
Number of households (million)	27.5	25	30	27.5	27.5
Energy use per household	Large reduction	Very large reduction	Small reduction	Similar to current	Similar to current
Supply mix	Coal (with and without CCS[11]); renewables; H_2; biofuels	Coal (with CCS); nuclear; CHP[11]; biofuels	Gas (with and without CCS); biofuels; nuclear; H_2; renewables	Nuclear; renewables H_2; biofuels	Nuclear; CCS (coal and gas); renewables; biofuels

Decarbonisation policies	Innovation and technology driven	Collectivist approaches to demand-side policy	Similar to 2002 with focus on supply	Strongly market-focussed government	Strongly market-focussed government
Transport	Low growth in aviation; reduction in car use; very large increase in public transport	Medium growth in aviation; low growth in car use; large increase in public transport	High growth in aviation; no growth in car use; small increase in public transport	V high growth in aviation; large growth in car use; large growth in public transport	V high growth in aviation; large growth in car use; large growth in public transport
Transport fuels	Oil; electricity; H_2	Oil; electricity; H_2	Oil; biofuels; electricity; H_2	Oil; biofuels; electricity; H_2	Oil; biofuels; electricity
Hydrogen	Stationary and transport uses; production from gasification with CCS and renewables; no pipelines	Transport uses; production from gasification with CCS, nuclear and renewables; no pipelines	All sectors including aviation; production from gasification with CCS, nuclear and renewables; pipelines and H_2 by wire	Stationary and transport uses; production from renewables and nuclear; extensive pipeline system	No hydrogen

Table 7.2 Aviation Summary Table for the Red Scenario

Annual average % growth in passenger-km	1.4%
Annual average % change in fuel efficiency	–1.6%
Energy consumption in 2050[12]	8.5 Mtoe
Carbon emissions in 2050[13]	7.5 MtC
Aviation fuel	Kerosene

The Turquoise Scenario

Of all five scenarios, the *Turquoise Scenario* has levels of energy consumption and economic growth most similar to those in 2002. The economy in 2050 is around three and a half times bigger, with a corresponding energy consumption of around 200Mtoe compared with 170Mtoe in 2002. The economy is diverse, with more dominant sectors including the commercial, construction and public sectors.

Demand-Side Characteristics Despite a higher overall energy consumption, the medium rates of economic growth require energy efficiency improvements to play a significant role in reaching the 60% target and serve to reduce the nation's energy intensity by over 60% by 2050. Different tiers of governance along with a number of agencies and departments are responsible for delivering decarbonisation through encouraging, and in some cases enforcing, improved energy efficiency, energy security and low-carbon energy supply. Markets are used selectively, for example providing incentives for decarbonisation in construction, private vehicle transport and aviation. A number of sectors, including public transport, are re-nationalised. The land transport modes showing growth are actually a reverse of those growing in 2002, and this is brought about through prioritisation of public transport over private cars.

Table 7.3 Aviation Summary Table for the Blue Scenario

Annual average % growth in passenger-km	3.7%
Annual average % change in fuel efficiency	–1%
Energy consumption in 2050 (10.5 Mtoe in 2002)	34 Mtoe
Carbon emissions in 2050 (9 MtC in 2002)	30 MtC
Aviation fuel	Kerosene

Supply-Side Characteristics The public sector takes on a dominant role in commissioning and planning new energy supply. By 2020, hydrogen end-use technologies are well developed, licensed and fully commercialised and public concerns over the safety of hydrogen as a transport fuel have been addressed. There is a new nuclear build programme and a revised common agricultural policy boosting energy crop industries. Coal continues to be a significant player in the UK's energy supply with the development of a series of pilot carbon capture experiments to test the viability for both gas and coal-fired stations in the early part of the century.

Aviation Industry Passenger-km travelled by air are more than eight times greater in 2050 than in 2002 and this, along with an increase in rail transport, implies a significant but manageable growth in infrastructure. Growth in passenger-km averages at 4.5% per year and by 2015, the decision has been made for large-scale, centralised planning to provide adequate rail and airport infrastructure. Taxes on aviation fuel combined with oil price rises have forced the airlines to form partnerships with the energy industry to drive low-carbon innovation within the aviation sector, producing the first commercial hydrogen-fuelled aircraft by 2030. Despite the widespread use of alternative fuels and energy efficiency improvements by 2050, aviation contributes the largest portion of the UK's CO_2 emissions in 2050 by some margin (see Figure 7.4). Table 7.4.

The Purple and Pink Scenarios

The *Purple* and *Pink Scenarios* both exhibit the same demand characteristics, but have different supply portfolios to demonstrate the range of supply options to meet the 60% carbon-reduction target in a high energy consumption UK. Both energy consumption and growth are high with the 2050 economy over six times larger than the 2002 economy and energy consumption approximately twice the baseline level. The commercial sector continues to dominate the economy but many of the UK's other industries continue to thrive.

Table 7.4 Aviation Summary Table for the Turquoise Scenario

Annual average % growth in passenger-km	4.5%
Annual average % change in fuel efficiency	–1.8%
Energy consumption in 2050 (10.5 Mtoe in 2002)	54 Mtoe
Carbon emissions in 2050 (9 MtC in 2002)	25 MtC
Aviation fuel	Kerosene; Hydrogen; Biofuel

Demand-Side Characteristics The UK's economic success is attributable to a vibrant and innovative market economy with a relatively small but supportive market-oriented government, all operating within a stable, highly globalised economy. The drive towards a low-carbon society arises from two fronts; an international obligation to significantly cut carbon emissions by 2050 and an increasing concern within energy markets over the insecurity associated with a reliance on imported fossil fuels. Demand for passenger transport has grown across all sectors with an overall sixfold increase in passenger-km travelled. These high levels of demand require large-scale developments in transport infrastructure. Although within high energy demand scenarios, substantial amounts of decarbonisation are required from low-carbon supply, and significant improvements in energy efficiency are also essential.

Supply-Side Characteristics—Purple Scenario The reluctance of the aviation industry to make step changes in fuel technology within the early part of the millennium leads to airlines having a heavy reliance on fossil fuels. Consequently, to reach the 60% carbon-reduction target, all of the other sectors have had to virtually completely decarbonise leading to a 'carbon-free' national grid dominated by nuclear and renewable energy. For heat and motive power, hydrogen and biofuels are widely available, and a mixture of combined heat and power schemes spread nationwide. Small on-site renewable technologies are also extremely popular.

Supply-Side Characteristics—Pink Scenario In this market-led society, the dominant fossil fuel companies reject the idea of a hydrogen economy due to the slow pace of research and development and instead invest heavily in carbon capture and storage (CCS) for electricity production. By 2010, a public–private partnership leads to an industry-led public awareness campaign about CCS in conjunction with a boost in privately funded university research. The construction of a new major gas pipeline from Russia is also complete and by 2030 the fossil fuel industry is booming with coal imports at an all-time high.

Aviation Industry There is a fourfold increase in domestic aviation and a tenfold increase in international aviation. Policies are put in place to encourage infrastructure developments with the building of many new airports and runways, particularly around Britain's coastline. Given the high demand for aviation in both scenarios, alternatives to kerosene are required well before 2050. Although hydrogen as a fuel has begun to penetrate the land transport sectors within the *Purple Scenario*, the required technological developments have not been adequate to develop hydrogen for commercial use within the aviation industry. Biofuels, however, have made small inroads into addressing the carbon intensity of the aviation industry. In the *Pink Scenario*, biofuels are considered the most viable low-carbon fuel and are used widely by 2040, but again hydrogen-fuelled aircraft do not exist. Despite the changes to aviation fuel, Figure 7.4 demonstrates that if the industry grows at 5%

per year on average between 2002 and 2050, then this sector will be by far the dominant carbon-emitting sector. Table 7.5.

Scenario Discussion

A clear message emerges from the scenarios in relation to the important sectors in 2050; namely that the household, industry and aviation Figure 7.2) are the energy-dominant sectors. By contrast, a number of sectors, even when experiencing high growth rates, remain relatively low-consuming sectors, namely public administration and public transport. Public transport is particularly important as, in terms of passenger-km, it can absorb significant modal shifts from higher energy consuming transport sectors. For example, a slightly higher energy consumption associated with a significant increase in public transport will be more than compensated by the very substantial reduction in energy consumed as passengers substitute private cars for the train, tram and bus.

An overview of the energy consumption and CO_2 emission figures for the scenarios is presented in Table 7.6 and Table 7.7. The total primary demand figure takes account of the transformation of fuel into electricity, the energy industry's own energy use and losses within the system, hence it is considerably higher than the 'Total consumption' figure. However, the CO_2 emissions associated with transformation in, for example, household electricity, is taken into account in the household CO_2 figure.

It is apparent from the tables that the sectors with the highest energy consumption are not necessarily the sectors emitting the most CO_2. This is due to the variations in energy supply. A summary of the supply fuels for 2050 is presented in Figure 7.3. These integrated scenarios encompass a wide range of supply portfolios including different electricity generating options, as well as heat and motive fuels from sources both in wide use today, and those that will require new technology or policies for their implementation. Although all scenarios have a suite of electricity supply technologies, the higher the energy consumption, the lower the flexibility in terms of fuel choice. For

Table 7.5 Aviation Summary Table for the Purple and Pink Scenarios

Annual average % growth in passenger-km	5%
Annual average % change in fuel efficiency	–1%
Energy consumption in 2050 (10.5 Mtoe in 2002)	65 Mtoe
Carbon emissions in 2050 (9 MtC in 2002)	44 MtC (Purple) 39 MtC (Pink)
Aviation fuels	Kerosene (Pink/Purple) Biofuel (Pink/Purple)

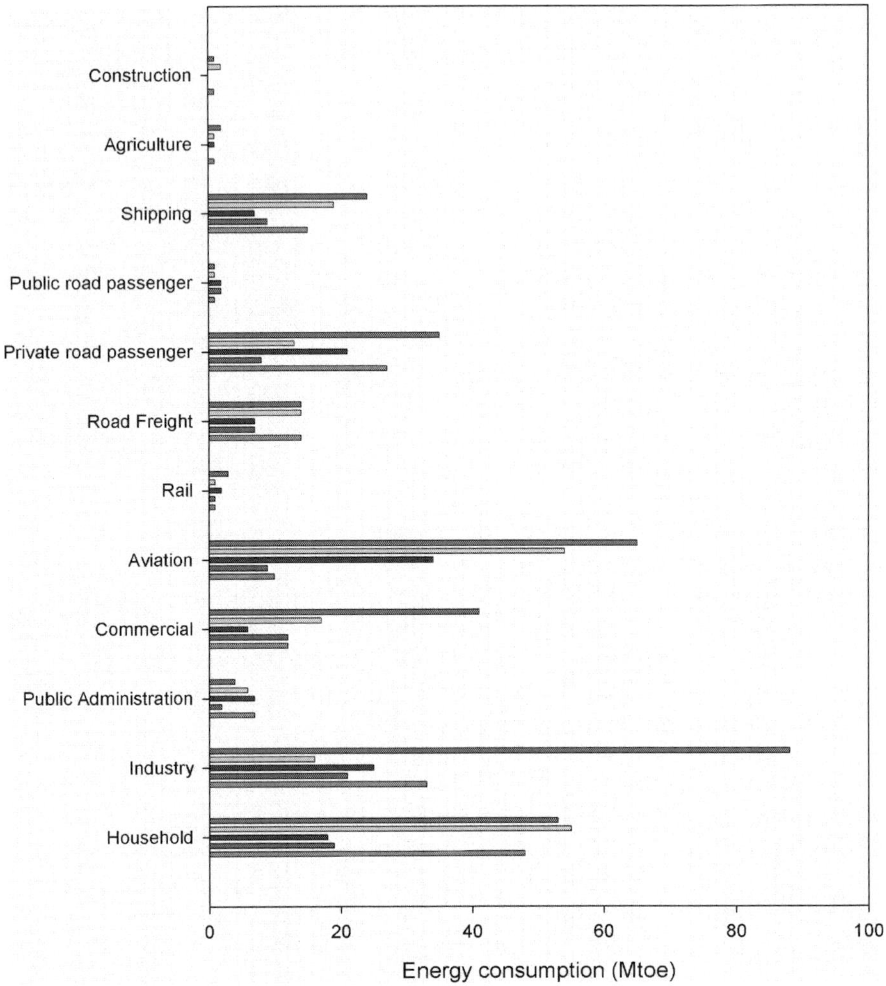

Figure 7.2 Energy consumption for the Tyndall 60% decarbonisation scenarios. The top bar for each sector represents the Purple/Pink scenarios, next Turquoise, next Blue, next Red and finally Orange (baseline). Therefore, the bar closest to the bottom of the graph is Orange.

example, the *Purple* and *Pink Scenarios* rely on widespread large supply-side options such as coal and gas with carbon capture and storage, nuclear power and renewable technologies. As such, high-energy scenarios are more likely to face greater planning, development and regulation hurdles than low-energy futures.

Figure 7.3 illustrates that, in all of the scenarios, oil remains an important transport fuel, particularly for the aviation industry. With few opportunities for large-scale use of alternative low-carbon transport fuels for aviation over the timescale considered, compared with a plethora of options for the other

Table 7.6 Scenario Baseline and Endpoint (2050) Energy Summary

	Energy Consumption (Mtoe)				
	2002	*2050*			
Sector	*Baseline*	*Red*	*Blue*	*Turquoise*	*Purple/Pink*
Household	48	19	18	55	53
Industry	51	22	15	19	89
Service industries	20	24	14	24	47
Aviation	10	9	34	54	65
Other transport	58	27	39	48	76
Total Transport	68	36	73	102	141
Total consumption	170	91	130	199	330
Total primary demand	243	140	191	304	475/505[14]

sectors, the aviation industry consumes an increasing portion of the carbon cake as energy consumption increases with each scenario (Figure 7.4).

For the UK Government's original 60% target, which ignores the impact of cumulative emissions, it is possible to reconcile high-energy consumption

Table 7.7 Scenario CO_2 Emission Summary

	CO_2 Emissions (MtC)					
	2002	*2050*				
Sector	*Baseline*	*Red*	*Blue*	*Turquoise*	*Purple*	*Pink*
Household	40	12	8	16	1	3
Industry	32	15	12	5	2	8
Service industries	19	9	5	7	2	2
Aviation	9	8	30	25	44	40
Other transport	48	17	6	7	12	7
Total Transport	57	25	35	33	56	47
Total consumption	148	60	61	60	61	60
Total primary demand	162	65	66	65	65	65

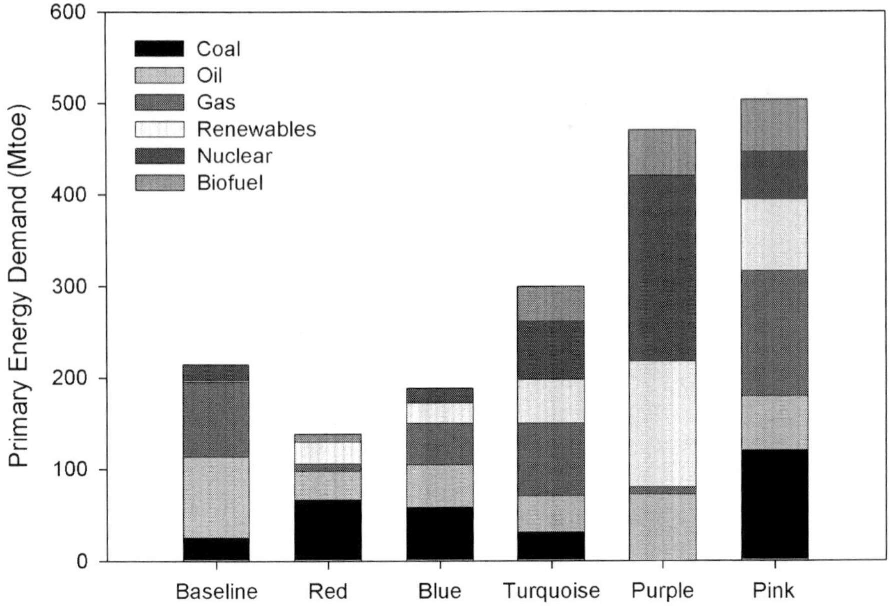

Figure 7.3 Primary energy demand for Tyndall's first set of Integrated Scenarios.

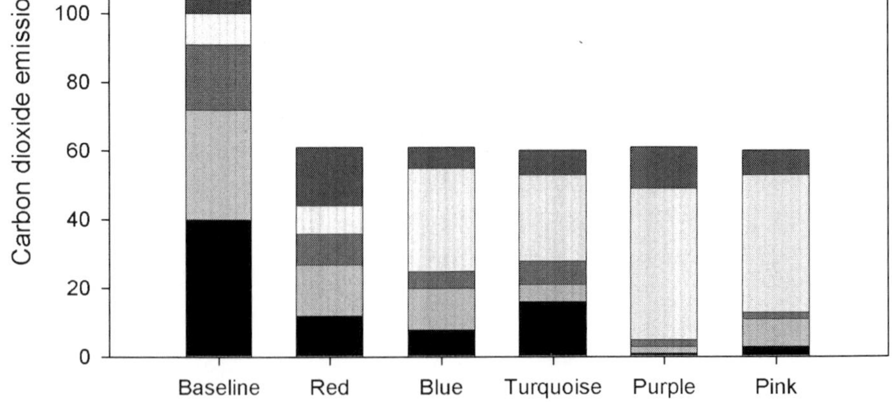

Figure 7.4 CO_2 emissions for each cumulative carbon scenario.

with a low-carbon future, albeit demanding significantly more innovative management structures and policies to drive technological and behavioural change. However, stakeholders involved in the scenario process found the high energy supply scenarios to be significantly more challenging in terms of sustainability due to the substantial infrastructure changes required.

The 60% energy scenarios presented here demonstrate some of the implications for other sectors of the UK's economy of continued growth within the aviation industry. Even if aviation is permitted to grow at rates of even half those being seen today, by 2050 CO_2 emissions from the aviation sector will far exceed emissions from any other sector of the UK's economy. For the UK to reach a 60% carbon-reduction target by 2050 (65MtC), all of the other sectors of the economy will have had to make substantial cuts to their emissions—over and above 60%—in addition to the aviation industry incorporating some significant technological developments. This is not to say this is not feasible, but certainly extremely challenging. However, in the context of a cumulative 450 ppmv CO_2 pathway and budget, the challenge becomes significantly more demanding, as illustrated in the next section.

TYNDALL'S CUMULATIVE CARBON SCENARIOS

In contrast to the first set of Tyndall Integrated Scenarios, the second set focuses primarily on exploring mobility, with more attention paid to the detail of passenger transport. In this case, scenarios exploring contrasting mobility patterns were developed within a number of energy-system boundaries such as the CO_2 pathway commensurate with the 450 ppmv CO_2 stabilisation level, in addition to constraints on the availability of biofuels and nuclear power.[15] The combination of the energy-system boundaries, the stringent CO_2 emission pathway and an aim to investigate differences in mobility resulted in two scenarios that have many similarities within the non-mobility sectors. Differentiation in the supply-side emerged as a consequence of differences in their patterns of demand for transport.

Taking a similar approach to Tyndall's 60% energy scenarios, these scenarios do not explore *where we will be* if certain trends and drivers are projected forward, but instead, *where we have to be* if we are to remain within the carbon budget. No judgements are made as to the desirability or likelihood of the scenarios, with the aim being to provide a neutral context for the policy framework by which the carbon reductions can be achieved.

Scenario Descriptions

The two scenarios developed are entitled *Static Mobility* and *Mobility Plus*. Both have medium economic growth and low energy demand by 2050. With rates of economic growth similar to today, the economy by 2050 is three times bigger than in 2004 (the baseline year) with an associated energy consumption that has been reduced to half of current levels. The UK remains a service economy driven by the commercial and public administrative sectors. The productive sectors collectively contribute the remaining 14% of UK GDP, primarily through industry and construction.

The *Static Mobility* scenario is characterised by a ceasing of growth in passenger transport, so that the same numbers of passenger-km are travelled in 2050 as 2004. By contrast, in the *Mobility Plus* scenario, growth in mobility is maintained, though at a lower rate, with UK individuals travelling twice as many passenger-km by road and rail in 2050 and three times more passenger-km by air. The final energy demand for all sectors by 2050 differs between the two scenarios by 16Mtoe. The majority of this difference is attributable to the additional energy required for the increase in passenger-km in the *Mobility Plus* scenario. Figure 7.5 and Figure 7.6 illustrate the final energy demand in each broad sectoral group for the two scenarios over the forty-five-year time frame and Table 7.8 presents a summary of the key scenario parameters.

By 2030 the carbon pathway, outlined in Chapter 3, requires significant reductions to be made in both energy consumption and CO_2 emissions. Within the construction sector, new buildings must be constructed to the highest standards of energy efficiency, and similarly all retrofits and refurbishments must significantly reduce CO_2 emissions by improving energy

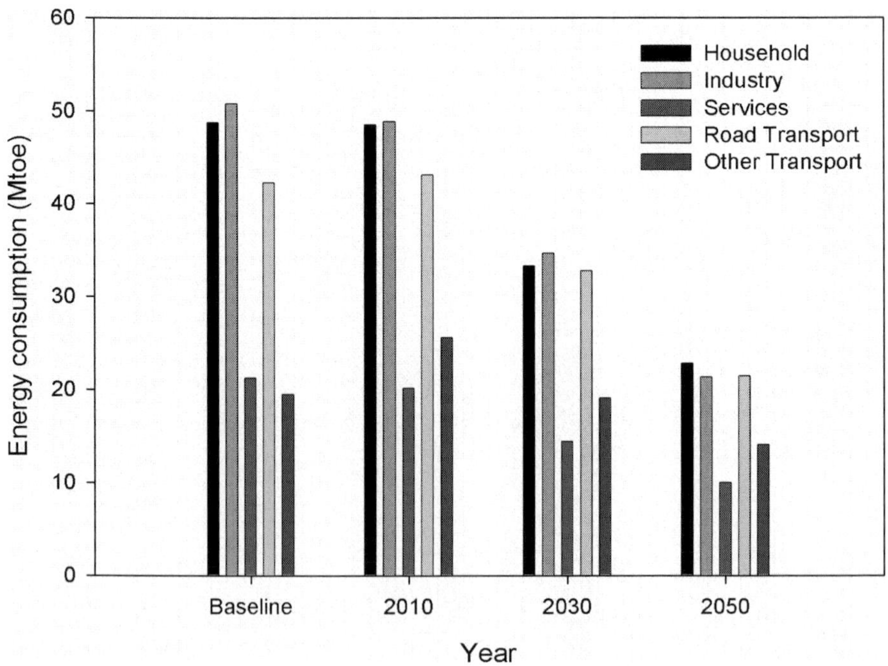

Figure 7.5 Final energy demand for the *Static Mobility* scenario in the baseline year and endpoint years of 2010, 2030 and 2050.

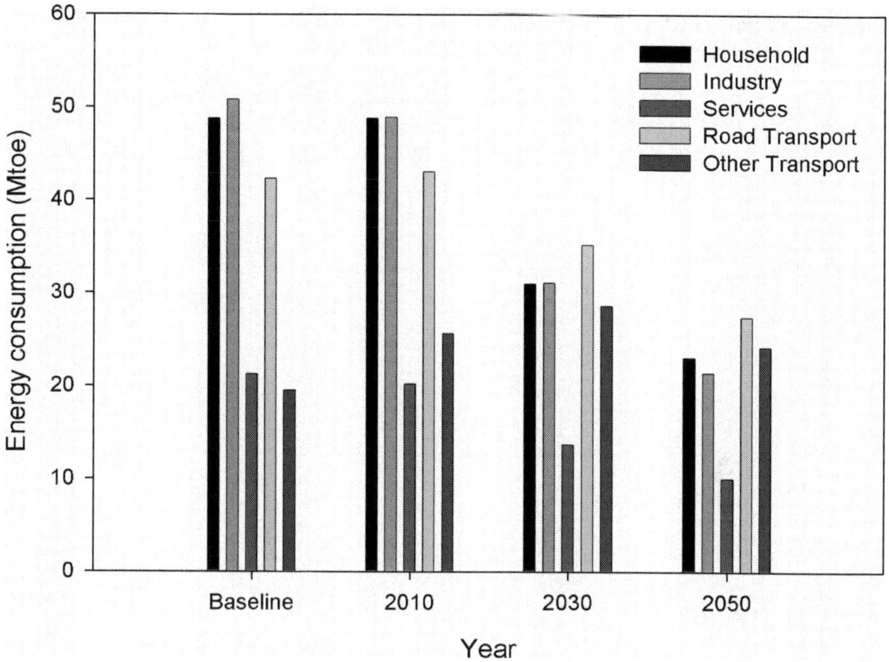

Figure 7.6 Final energy demand for the *Mobility Plus* scenario in the baseline year and endpoint years of 2010, 2030 and 2050.

consumption and incorporating low-carbon supply technologies. In terms of differences between these scenarios' energy demand portfolios, a larger portion of final energy demand is consumed by the household and industry sectors than by transport in the *Static Mobility* scenario, whereas in the *Mobility Plus* scenario, transport constitutes the largest portion of final energy demand (Figure 7.5 and 7.6).

By 2050, energy consumption in both scenarios has been significantly reduced across almost all sectors of the economy, primarily through increases in the technical efficiency of delivering goods and services and by changes to consumption habits. Despite the overall large-scale reduction in energy consumption, keeping within the stringent CO_2 pathway has, in addition, necessitated considerable decarbonisation of the supply system, and a shift to the use of hydrogen as an energy carrier. Given these scenarios focussed on passenger transport, this broad sector will be described in a little more detail, with a particular focus on aviation.

Transport

In the baseline year of 2004, private road transport accounted for the largest proportion of energy consumption of all of the transport sectors, double that of aviation, and second only to the household sector. In terms of energy

Table 7.8 2050 Scenario Summary

		Baseline (2004)	*Static Mobility*	*Mobility Plus*
Final energy demand (Mtoe)		183	90	106
Primary energy demand (Mtoe)		237	118	138
Primary Fuel Proportion	Renewables	1%	32%	29%
	Fossil fuels with CCS	0%	40%	42%
	Conventional fossil fuels	89%	17%	15%
	Biomass	2%	11%	14%
	Nuclear	8%	0%	0%
	Electricity proportion: grid	94%	75%	78%
	Electricity proportion: decentralised	6%	35%	33%
Transport	Passenger-km: road	736	717	1379
	Passenger-km: rail	51	71	199
	Passenger-km: air	273	271	793
	Occupancy: car	1.6	1.8	1.7
	Occupancy: rail	93	120	130
	Occupancy: air international	177	200	250
Carbon Emissions (MtC)	Land transport	34.7	2.7	1.4
	Air transport	9.75	3.6	6.08
	Industry	48.6	3.5	3.9
	Services	22.3	1.4	1.1
	Households	42	4.3	4.1
Energy efficiency: households (ttoe[16]/household)		1.93	0.82	0.83
Hydrogen demand (Mtoe)		None	26	29
Total energy carbon emissions		164	17	17

efficiency, cars have been improving over recent years at a rate of around 1.7% per passenger-km per year. However, very recently, fuel efficiency declined for the first time in many years. This is thought to be the result of a boom in larger, heavier vehicles which traditionally exhibit poor fuel efficiency. Although accounting for some 6% of the UK's emissions in 2006, the aviation industry is growing at the highest rate of all of the sectors in the economy and showing little improvement in the way of fuel efficiency when considering UK data.

Within the two scenarios, the aviation sector develops somewhat differently, with the changes being more apparent in the later years. Between 2004 and 2010, the recent high rates of growth are assumed to continue, fuelled by the ongoing boom in low-cost airlines and the attempts to mimic low-cost models by more traditional airlines. In later years, and in order to remain within the 450 ppmv carbon pathway, aviation is required to employ a combination of demand management measures, improvements in energy efficiency through technological advances and operational change, and, particularly in the case of *Mobility Plus*, the use of alternative low-carbon fuels. Measures concerned with the aviation industry within these 'what-if' scenarios are summarised under the categories of demand management, fuel efficiency, alternative fuels and infrastructure.

Demand Management

Demand management is critical within both scenarios and, by 2030, growth rates in both domestic and international aviation have been forced to decline to 2.5% per year within *Static Mobility* resulting in a real-terms decrease in passenger-km travelled by domestic aviation relative to the baseline and a fractional decrease for international aviation by 2050. Modal shift to rail transport accounts for much of the decline in domestic aviation, and for a small amount of the passenger-km from international flights, particularly within Europe. In the early years of *Static Mobility*, the aviation industry focussed on improving fuel efficiency alone rather than carbon efficiency through alternative fuels, therefore to remain within budget, the industry was left with no choice but to curtail emissions through reducing growth in the medium-term. The reduction in growth reflects the beginning of a necessary shift of some of the domestic aviation passengers onto the rail and road networks, required under these carbon constraints.

Within *Mobility Plus*, growth rates reduce to 3.5% per year on average between 2010 and 2020, and down to 2% per year between 2020 and 2030, despite fuel efficiency improvements and the use of some alternative fuels. Domestic aviation passenger-km are the same in 2050 as they were in 2004, but overall the industry within this scenario is significantly larger in 2050 than in 2004 despite demand management measures curbing growth to less than 1% per year between 2030 and 2050. The industry has not only needed to improve fuel efficiency and reduce passenger-km growth to curtail emissions growth, but has also needed to improve carbon efficiency through the use of alternative fuels, particularly within *Mobility Plus*.[17]

Efficiency Improvements

Within *Static Mobility* and *Mobility Plus*, recent energy efficiency trends have been improved both in domestic and international aviation. The urgency and importance placed on climate change has persuaded the aviation industry to incorporate some additional non-technological measures, previously found to be unacceptable. The managerial changes required include, for example, new ticketing arrangements to ensure that plane load factors are greatly increased in relation to the baseline year. The proposed improvements to the air traffic management system highlighted by the industry in 2005[18] are implemented by 2020 to deliver the one-off improvement in fuel efficiency enabled by more direct flights, less use of holding patterns, less taxiing and fewer delays on runways.

Aircraft design features focus much more prominently on improving fuel efficiency. Unfamiliar airframe designs incorporating open-rotor engines, laminar flow and lighter carbon composite materials, for example, are used more widely. As a result of these technological and managerial changes, fuel efficiency improvements reach a maximum of 2.2% per year during the 2020s for *Static Mobility*, and 2.1% for *Mobility Plus*. By comparison, efficiency improvements up to 2.8% per year are assumed in private road transport in *Static Mobility*, and 4% per year in *Mobility Plus*. However, given that the rate of growth for car transport has recently been around 1% per year, the efficiency improvements offset the growth in passenger-km from car transport early on in both scenarios.

Fuels

In terms of energy supply in 2030, road transport is assumed to be using biofuel, oil, hydrogen and electricity, rail using similar fuels to those of today, and the aviation sector using some bio-kerosene and biodiesel. The two scenarios differ in terms of their levels of mobility to keep within the carbon budget; hence *Mobility Plus* has many more planes fuelled by a low-carbon source than *Static Mobility*. By 2030, the attitude within the aviation industry has altered significantly in relation to alternative fuels, due to the national and European drive towards eliminating CO_2 emissions. Bio-kerosene and biodiesel are therefore widely used to comply with the nation's drive towards a low-carbon economy, contributing to 50% of aviation fuel by 2050 in *Mobility Plus*.

Infrastructure

Within *Static Mobility*, occupancy rates and load factors have increased for both domestic and international flights. This increase in load factor, coupled with the 28% increase in passenger-km, could theoretically be accommodated within existing airport infrastructure by increasing landing rates on average by around 17% at existing airports. If, on the other hand, landing rates remained at current levels, this increase in passenger-km would require around 16% more runway capacity.

It is assumed in *Mobility Plus* that the aviation industry has taken a slightly different route in terms of aircraft fleet, with a trend towards smaller, lighter aircraft for domestic use and higher capacity aircraft for long-haul flights. In this case, the larger aircraft are reserved for the long-haul flights by placing limitations on their use for medium and shorter journeys. Assuming that the average aircraft for international aviation has 15% more capacity, and that load factors are increased to 90%, no increase in landing rates on current runways would require a doubling of the number of runways in existence today. If, on the other hand, planes could land on average a third more frequently, then roughly 50% more runways would be required. Clearly, there are additional issues in relation to sustainability that are apparent even with this basic analysis.[19]

Transport Summary

To achieve the carbon reduction necessary for the UK to play its part in stabilising atmospheric CO_2 concentrations at around 450 ppmv taking the cumulative budget approach, both land and air transport must significantly decarbonise. In 2004, passenger road transport was responsible for around 22MtC, road freight for 13MtC and aviation around 10MtC—a total of 30% of total UK CO_2 emissions. By 2050, these figures are reduced to 0.5MtC (passenger road transport), 1MtC (road freight) and 3.5MtC (aviation) in the *Static Mobility* scenario, and 0MtC (passenger road transport), 0MtC (road freight) and 6MtC (aviation) in *Mobility Plus*. No one measure is responsible for the reductions, but rather a comprehensive package of demand management, large incremental improvements in vehicle efficiency and a new low-carbon fuel chain have combined to bring about the necessary change.

In particular, the aviation industry has had to take responsibility for its emissions by significantly reducing its growth rates. Fuel efficiency improvements at rates higher than those expected by the industry in 2004, coupled with the use of biofuels to replace some oil-based kerosene, are found to be necessary within the scenarios if the UK is to develop meaningful policy on the basis of the 2°C target.

8 Conclusion

This book provides an overview of the aviation industry in the context of carbon mitigation commensurate with the EU and UK's commitment to making their fair contribution to avoiding 'dangerous climate change'. The book synthesises science and policy to explain why the aviation sector has become the focus of academics, environmental campaigners and increasingly policymakers. If industrialised nations are to play their part in ensuring global mean temperatures do not exceed a 2°C temperature rise above pre-industrial levels, a sea change in policies and actions to both reduce energy consumption and decarbonise energy supply is essential. Key to achieving an *absolute* reduction in emissions in the short- to medium-term is to accept that the target must include *all* sectors of the economy, and then to find the most appropriate and effective method for reducing emissions for the economy as a whole. Incorporating the aviation industry within the EU's Emissions Trading Scheme may go some way to address aviation's impact, but as implementation and significant emission reduction across many sectors is unlikely for some time, alternative measures are required now to ensure CO_2 emissions start to fall as a matter of urgency. Given that the aviation industry has significantly fewer technological and managerial options to reduce CO_2 emissions in the short- to medium-term than do most other sectors, and that any improvements made will likely be outstripped by growth, two clear courses of action emerge:

Industrialised nations must urgently reduce energy consumption, improve energy efficiency and roll out low-carbon supply options across *all* sectors of the economy. Only a portfolio of all of these measures will deliver the scale of decarbonisation required.

Industrialised nations must curb growth and stimulate innovation within the aviation industry until the sector's annual growth in CO_2 emissions can be offset by their annual improvements in fuel efficiency or ultimately carbon efficiency.

If industrialised nations do not begin to follow these two lines of action immediately and in earnest, the UK and other EU governments will miss their 2050 targets and more significantly will forgo any real opportunity to stabilise emission at levels appropriate to 2°C.

This book has primarily focused on CO_2 emissions from aircraft. However, it should be borne in mind that the aviation industry also contributes to climate change through emissions of soot and other aerosols, water vapour and NO_x. Addressing the full range of aircraft emissions is complicated, as highlighted in Chapter 2, but the overall conclusion remains the same. If growth rates within the aviation industry are to be curbed, this would curb the growth of all emissions. Furthermore, the additional aviation-induced warming must be considered in the light of the stark conclusion that all of the other sectors in industrialised economies must virtually decarbonise to permit aviation to grow at rates well in excess of GDP growth. The challenge to climate policy posed by aviation and compounded by the non-CO_2 emissions leads to the inescapable conclusion that governments of industrialised nations must act now to curb aviation growth. This need not mean stasis or decline in passenger-km within these nations, but responsible aviation growth must be in the context of urgent and rapid reductions in carbon emissions.

Notes

NOTES TO CHAPTER 1

1. (Keynote 2007)
2. (National Bureau of Statistics China 2006)
3. (IPCC 2007; page 12 of the summary for policymakers)
4. (Jones et al. 2006)
5. (Penner et al. 1999)
6. (Anderson and Bows 2007)

NOTES TO CHAPTER 2

1. (RCEP 2002, 13)
2. One RPK is one paying passenger carried for one kilometre: the indicator reflects both the number of passengers carried and the distance they are carried.
3. (Penner et al. 1999, 3)
4. (Airbus 2007; Boeing 2007)
5. (World Bank 2006)
6. (Penner et al. 1999)
7. (De La Fuente Layos 2007)
8. Passenger statistics for Poland do not extend back to 1993.
9. http://unfccc.int/ghg_emissions_data/items/3800.php
10. Passengers passing through UK airports.
11. (CAA 2006)
12. (DTI 2005; energy balances)
13. (Gordon et al. 2005)
14. (Boeing 2007)
15. (Rolls-Royce 2007)
16. (EUROCONTROL 2007b; EUROCONTROL 2007c; EUROCONTROL 2007a)
17. Therefore taking into account traffic between rapidly growing markets such as those between the EU and Asia.
18. As will be discussed in Chapter 4.
19. (DfT 2004b)
20. (DfT 2004a)
21. (DfT 2003)
22. Capacity is inherently constrained in such forecasts by, for example, existing landing charges, car park fees etc.

23. See paragraph 3.53 within the DfT Aviation and Global Warming paper (DfT 2004a).
24. (Sausen et al. 2005)
25. The original IPCC Special Report on Aviation had made a best estimate of 2.7 × carbon dioxide alone.
26. Partnership for Air Transportation Noise and Emissions Reduction, 2006, report of the workshop on the Impacts of Aviation on Climate Change. Cambridge, MA, June 2006.
27. (Williams and Noland 2005)
28. The Quantify project: http://www.pa.op.dlr.de/quantify/.
29. (Nakicenovic et al. 2000)
30. (CDIAC 2007)
31. (IEA 2006; CDIAC 2007)
32. (UNFCCC 2006)

NOTES TO CHAPTER 3

1. (COMM 2007, 2)
2. (DEFRA 2006)
3. (DTI 2006)
4. (DTI 2003)
5. (IPCC 2007)
6. (Nakicenovic et al. 2000)
7. In terms of temperatures in 2090–2099 relative to 1980–1999 (IPCC 2007, Table SMP.3).
8. (Raupach et al. 2007)
9. (Jones et al. 2006)
10. (CDIAC 2007)
11. Measurements of carbon dioxide emissions are either given in tonnes of CO_2 or tonnes of Carbon (C). 1 tonne of CO_2 is 3.67 times 1 tonne of carbon. Here the square brackets denote tonnes of carbon.
12. (Meinshausen 2006)
13. See http://www.realclimate.org/index.php/archives/2007/10/co2-equivalents/ for a discussion of this issue.
14. Organisation for Economic Co-operation and Development—group representing the interests of industrialised nations.
15. This is also the case for international shipping activities and related emissions.
16. (IPCC 2007, 2)
17. (Anderson and Bows 2007)
18. (Warren 2006)
19. (Cranmer et al. 2001; Cox et al. 2006)
20. (Matthews 2005; Jones et al. 2006; Matthews 2006)
21. (Bows et al. 2006; Stern 2006)
22. (Bows et al. 2006, 20)
23. (Meyer 2000)
24. (Hohne et al. 2003)
25. For another accessible account of post-Kyoto options, see: www.fiacc.net/app/approachlist.htm.
26. This does not assume the existence of the Clean Development Mechanism, itself a political compromise. If this mechanism is assumed, this objection to Contraction & Convergence can be ignored.
27. (Bows et al. 2005)

28. (IPCC 2007, 16; summary for policymakers)
29. (UNFCCC 1997)
30. (Stern 2006)
31. (Bows et al. 2006)

NOTES TO CHAPTER 4

1. This concern, particularly in relation to the opportunity cost of curtailing the life span of currently new aircraft in favour of more efficient designs, is discussed in Greener by Design's annual report 2006–2007 (Greener by Design 2007).
2. (Green 2005a; Cairns and Newson 2006)
3. (Bows et al. 2005; Bows et al. 2006a; Bows and Anderson 2007)
4. (RCEP 2002)
5. (Greener by Design 2007)
6. (Windischbauer and Richardson 2005)
7. (Anderson and Wood 2001)
8. See http://www.boeing.com/news/frontiers/archive/2002/september/i_pw.html.
9. (Saynor et al. 2003)
10. (Ponater et al. 2003)
11. (Penner et al. 1999)
12. (Gauss et al. 2003)
13. (Little 2000)
14. (Goldblatt et al. 2005)
15. (Mannstein et al. 2005)
16. (Williams et al. 2003; Fichter et al. 2004)
17. (Boeing 2007)

NOTES TO CHAPTER 5

1. (UNFCCC 1997)
2. (RCEP 2000, section 2.24)
3. Their strategic objectives are outlined on their web site www.icao.int.
4. (Penner et al. 1999)
5. See www.icao.int/icao/en/env/a35-5.pdf
6. (Friedlingstein et al. 2001)
7. This figure is closer to 40% when considering the total to include international aviation and shipping emissions.
8. (Wit et al. 2005)
9. (Mannstein et al. 2005)
10. (British Airways 2007)
11. (Anderson et al. 2007)
12. (Ouzky 2007)
13. Based on Table 4 within Tyndall's report on the EU ETS, *Aviation in a Low-Carbon EU* (Anderson et al. 2007).
14. Theoretically, multiple instruments are likely suboptimal in terms of economic efficiency. In practice, this may not be a problem.
15. (COMM 1999)
16. (Wit et al. 2002)
17. (Janic 2003)
18. (DETR 1998)
19. (DfT 2003)

20. (DfT 2004b)
21. (CAA 2006)
22. (De La Fuente Layos 2007)
23. (DTI 2003)
24. (Bows and Anderson 2007)
25. (DfT 2004b)
26. ACARE assume 50% fuel efficiency improvement between 2000 and 2050. To incorporate this, 15% is assumed to be between 2000 and 2030, with a further 25% occurring between 2030 and 2050. The remaining 10% is already factored into the original DfT figures and arises from assumed improvement in operational measures in aviation.
27. (House of Lords 2004)
28. (DEFRA 2007)
29. (Anderson and Bows 2007)
30. For more information on the impact of including all arrivals and departures within the EU ETS, see Anderson and Bows 2007.

NOTES TO CHAPTER 6

1. (DfT 2005)
2. (DTI 2005)
3. (Pongas 2000; Tronet 2004)
4. (Tronet 2002)
5. The low-cost airlines have disrupted the notion of maturity in this context, but it can be bluntly interpreted as referring to a state of reduced or modest potential for growth.
6. (Lim et al. 2005)
7. (DfT 2003; DfT 2004a)
8. Eurostat data sources are available to subscribers from www.esds.ac.uk or online through the Eurostat web site at http://epp.eurostat.ec.europa.eu/portal/page?_pageid=1090,30070682,1090_33076576&_dad=portal&_schema=PORTAL
9. (EUROCONTROL 2007a)
10. (Anderson et al. 2007)
11. (Meyer 2000; this is the same apportionment regime used to produce the 60% UK carbon-reduction target)
12. Since there are already three scenarios from 2006 to 2012, this will result in nine scenarios overall in line with a 450 ppmv future, and three additional scenarios outside of this remit.
13. (Penner et al. 1999, Section 7.2.4)
14. (Green 2005a)
15. 'Op & tech' refer to operational and technological improvement.
16. This figure is assumed to be based on ongoing technical improvements in the efficiency of the fleet mean combined with a gradual shift towards low/zero-carbon fuels.
17. With less of an incentive to combat climate change, it is assumed that progress towards both better efficiency and low-carbon fuels is slower than in the other scenarios.
18. (Bows et al. 2006a)
19. This figure is slightly different from the CAA value of 7% shown in figure as it includes the value for 2001 rather than averaging over 1993–2000.
20. (DTI 2005)
21. (Upham 2003)

NOTES TO CHAPTER 7

1. (Anderson 2005)
2. (Anderson et al. 2005; Anderson et al. 2006; Bows et al. 2006; Mander et al. 2007)
3. (PIU 2002; DTI 2005)
4. The twin-axis framework essentially provides a two-dimensional approach, which considers, for example, community to be polar opposite of consumerism, or autonomy polar opposite of interdependence.
5. (Anderson 2001)
6. (Berkhout et al. 2002)
7. (Lovins 1977; Robinson 1982; Anderson 2001)
8. www.tyndall.ac.uk and www.tyndall.manchester.ac.uk
9. (DTI 2003)
10. First published in Anderson 2001.
11. CCS is carbon capture and storage; CHP is combined heat and power.
12. 10.5Mtoe in 2002.
13. 9MtC in 2002.
14. The *Purple* and *Pink Scenarios'* primary energy demand figures differ due to different supply mixes requiring differing amounts of transformation, distribution and losses.
15. The development of these scenarios was funded by Friends of the Earth and the Co-op Bank; therefore some of the energy-system constraints were consistent with their assessments of *sustainable* low-carbon supply options as opposed to using all feasibly available low-carbon supplies, as was the case with the first set of Tyndall Integrated Scenarios.
16. The abbreviation ttoe is thousand tonnes of oil equivalent.
17. Although some stakeholders within the industry indicated that alternative fuels would not be used within the aviation sector prior to 2030, Fischer-Tropsch kerosene produced from biofuel and biodiesel are both considered viable in today's aircraft with some marginal improvements and research and development (Sausen et al. 2005). Even if this is a kerosene extender rather than 100% blends, conventional biodiesel can be prone to solidify at low temperatures.
18. (Aviation 2005)
19. This is based on there being forty-seven runways in 2004.

Bibliography

Airbus. 2007. "Global Market Forecast: The Future of Flying." Global Market Forecast. http://www.airbus.com/en/corporate/gmf/.

Anderson, K. L. 2001. "Reconciling the Electricity Industry with Sustainable Development: Backcasting—A Strategic Alternative." *Futures* 33 (7): 607–623.

Anderson, K., and A. Bows. 2007. "A Response to the Draft Climate Change Bill's Carbon Reduction Targets." Tyndall Centre Briefing Note, Tyndall Centre for Climate Change Research. http://www.tyndall.ac.uk/publications/briefing_notes/bn17.pdf.

Anderson, K., A. Bows, and A. Foottit. 2007. *Aviation in a Low-Carbon EU*. Report for Friends of the Earth. Manchester: The Tyndall Centre for Climate Change Research.

Anderson, K., A. Bows, S. Mander, S. Shackley, P. Agnolucci, and P. Ekins. 2006. *Decarbonising Modern Societies: Integrated Scenarios Process and Workshops*. Tyndall Centre Technical Report. Manchester: The Tyndall Centre for Climate Change Research.

Anderson, K., S. Shackley, S. Mander, and A. Bows. 2005. *Decarbonising the UK: Energy for a Climate Conscious Future*. Manchester: The Tyndall Centre for Climate Change Research.

Anderson, K., and L. Wood. 2001. "Airships as a Means of Freight Transport, A Feasibility Study." CATE, REF: GR/R 32499.

Aviation. 2005. *A Strategy Towards Sustainable Development of UK Aviation*. Sustainable Aviation.

Berkhout, F., J. Hertin, and A. Jordan. 2002. "Socio-Economic Futures in Climate Change Impact Assessment: Using Scenarios as 'Learning Machines'." *Global Environmental Change* 12 (2): 83–95.

Boeing. 2007. "Current Market Outlook: How Will You Travel Through Life?" Boeing Market Forecasts, Boeing. http://www.boeing.com/commercial/cmo/pdf/Boeing_Current_Market_Outlook_2007.pdf.

Bows, A., and K. L. Anderson. 2007. "Policy Clash: Can Projected Aviation Growth be Reconciled with the UK Government's 60% Carbon-Reduction Target?" *Transport Policy* 14 (2): 103–110.

Bows, A., K. Anderson, and P. Upham. 2005. *Growth Scenarios for EU & UK Aviation: Contradictions with Climate Policy*. Manchester: The Tyndall Centre for Climate Change Research.

———. 2006a. *Contraction & Convergence: UK Carbon Emissions and the Implications for UK Air Traffic*. Tyndall Centre Technical Report. Norwich: The Tyndall Centre for Climate Change Research.

Bows, A., S. Mander, R. Starkey, M. Bleda, and K. Anderson. 2006. *Living Within a Carbon Budget.* Report commissioned by Friends of the Earth and the Co-operative Bank. Manchester: The Tyndall Centre for Climate Change Research.

British Airways. 2007. "Air Transport and Climate Change." http://www.britishairways.com/travel/crglobalwarm/public/en_gb (accessed October 2007).

CAA. 2006. "Main Outputs of UK Airports." http://www.caa.co.uk/default.aspx?catid=80&pagetype=88&sglid=3&fld=2006Annual (accessed April 1, 2007).

Cairns, S., and C. Newson. 2006. "Predict and Decide: The Potential of Economic Policy to Address Aviation-Related Climate Change." Demand reduction theme, UKERC, University of Oxford.

CDIAC. 2007. "Carbon Dioxide Emission Trends." http://cdiac.esd.ornl.gov/trends/emis/annex.htm (accessed July 2007).

COMM. 1999. *Air Transport and the Environment—Towards Meeting the Challenges of Sustainable Development.* Commission of the European Committees, Brussels, 640.

———. 2007. *Limiting Global Climate Change to 2 Degrees Celsius: The Way Ahead for 2020 and Beyond.* Final ed. Commission of the European Communities, Brussels.

Cox, P. M., C. Huntingford, and C. D. Jones. 2006. "Conditions for Sink-to-Source Transitions and Runaway Feedbacks from the Land Carbon-Cycle." In *Avoiding Dangerous Climate Change*, ed. H. J. Schellnhuber, W. Cramer, N. Nakicenovic, T. Wigley, and G. Yohe, 155–161. Cambridge: Cambridge University Press.

Cranmer, W., A. Bondeau, F. I. Woodward, I. C. Prentice, R. A. Betts, V. Brovkin, P. M. Cox, V. Fisher, J. A. Foley, A. D. Friend, C. Kucharik, M. R. Lomas, N. Ramankutty, S. Sitch, B. Smith, A. White, and C. Young-Molling. 2001. "Global Response of Terrestrial Ecosystem Structure and Function to CO2 and Climate Change: Results from Six Dynamic Global Vegetation Models." *Global Change Biology* 7:357–373.

DEFRA. 2006. *Climate Change: The UK Programme 2006.* UK Government Publication, Norwich.

———. 2007. *Draft Climate Change Bill.* Cm. 7040.

De La Fuente Layos, L. 2007. *Air Transport in Europe in 2005: Statistics in Focus.* European Communities, Brussels EU.

DETR. 1998. *A New Deal for Transport: Better for Everyone.* London Stationery Office: DETR.

DfT. 2003. *Aviation and the Environment: Using Economic Instruments.* Wetherby: Department for Transport.

———. 2004a. *Aviation and Global Warming.* The Stationery Office, London: Department for Transport.

———. 2004b. *The Future of Air Transport, Aviation White Paper.* HMSO, London: Department for Transport.

———. 2005. *Transport Statistics Great Britain: TSGB.* London: National Statistics.

DTI. 2003. *Our Energy Future—Creating a Low Carbon Economy, Energy White Paper.* The Stationery Office, London: Department of Trade and Industry.

———. 2005. *Digest of United Kingdom Energy Statistics.* The Stationery Office, London: Department of Trade and Industry.

———. 2006. *Our Energy Challenge: Securing Clean, Affordable Energy for the Long Term.* London: Department of Trade and Industry.

EUROCONTROL. 2007a. *Flight Movements 2006–2025. Long-Term Forecast.* EUROCONTROL, STATFOR, Brussels.

———. 2007b. *Flight Movements 2007–2008. Short-Term Forecast.* EUROCONTROL, STATFOR, Brussels.

———. 2007c. *IFR Flight Movements 2007–2013. Medium-Term Forecast.* EUROCONTROL, STATFOR, Brussels.

Fichter, C., S. Marquart, R. Sausen, and D. S. Lee. 2004. "The Impact of Cruise Altitude on Contrails and Related Radiative Forcing." *Meteorologische Zeitschrift*: 563–572.

Friedlingstein, P., L. Bopp, P. Ciais, J. Dufrense, L. Fairhead, H. LeTreut, P. Monfray, and J. Orr. 2001. "Positive Feedback Between Future Climate Change and the Carbon Cycle." *Geophysical Research Letters* 28:1543–1546.

Gauss, M., I. Isaksen, S. Wong, and W. C. Wang. 2003. "Impact of H2O Emissions from Cryoplanes and Kerosene Aircraft on the Atmosphere." *Journal of Geophysical Research* 108 (D10).

Goldblatt, D. L., C. Hartmann, and G. Durrenberger. 2005. "Combining Interviewing and Modeling for End-User Energy Conservation." *Energy Policy* 33 (2): 257–271.

Gordon, D. J., A. Blaza, and W. R. Sheate. 2005. "A Sustainability Risk Analysis of the Low Cost Airline Sector." *World Transport Policy & Practice* 11 (1): 13–34.

Green, J. E. 2005a. "Air Travel—Greener by Design. Mitigating the Environmental Impact of Aviation: Opportunities and Priorities." *The Aeronautical Journal* (September): 361–418.

———. 2005b. "Future Aircraft—Greener by Design?" *Meteorologische Zeitschrift*: 583–590.

Greener by Design. 2007. *Air Travel—Greener by Design Annual Report 2006–2007*. Royal Aeronautical Society, London.

Hohne, N., C. Galleguillos, K. Blok, J. Harnisch, and D. Phylipsen. 2003. *Evolution of Commitments Under the UNFCCC: Involving Newly Industrialised Economies and Developing Countries*. Berlin: Umweltbundesamt Research Report.

House of Lords. 2004. "The EU and Climate Change." *Lords Select Committee on the European Union (Sub-Committee E on Environment and Agriculture)*, 30. London: The Stationery Office.

IEA. 2006. "CO2 Emissions from Fuel Consumption." International Energy Agency. http://data.iea.org.

IPCC. 2007. "Climate Change 2007: The Physical Science Basis." Intergovernmental Panel on Climate Change, Contribution of Working Group I to the Fourth Assessment Report of the IPCC.

Janic, M. 2003. "The Potential for Modal Substitution." In *Towards Sustainable Aviation: Trends and Issues*, 131–148, ed. P. Upham, J. Maughan, D. Raper, and C. Thomas. London: Earthscan.

Jones, C. D., P. M. Cox, and C. Huntingford. 2006. "Impact of Climate–Carbon Cycle Feedbacks on Emissions Scenarios to Achieve Stabilisation." In *Avoiding Dangerous Climate Change*, ed. H. J. Schellnhuber, W. Cramer, N. Nakicenovic, T. Wigley, and G. Yohe, 323–331. Cambridge: Cambridge University Press.

Keynote. 2007. "Market Report 2007, Airports." Keynote reports, D. Fenn Middlesex.

Lim, L., D. Lee, and S. C. B. Raper. 2005. "The Role of Aviation Emissions in Climate Stabilisation Scenarios." Poster presented at *Avoiding Dangerous Climate Change*, Exeter.

Little, A. D. 2000. *Study into the Potential Impact of Changes in Technology of the Development of Air Transport in the UK*. Cambridge: DETR.

Lovins, A. 1977. *Soft Energy Paths*. London: Penguin.

Mander, S., A. Bows, K. Anderson, P. Agnolucci, S. Shackley, and P. Ekins. 2007. "Uncertainty and the Tyndall Decarbonisation Scenarios." *Global Environmental Change* 17 (1): 25–36.

Mannstein, H., P. Spichtinger, and K. Gierens. 2005. "A Note on How to Avoid Contrail Cirrus." *Transportation Research Part D: Transport and Environment* 10 (5): 421–426.

Matthews, H. D. 2005. "Decrease of Emissions Required to Stabilise Atmospheric CO2 Due to Positive Carbon Cycle–Climate Feedbacks." *Geophysical Research Letters* 32 (L21707): DOI:10.1029/2005GL023435.

———. 2006. "Emissions Targets for CO2 Stabilization as Modified by Carbon Cycle Feedbacks." *Tellus B* 58 (5): 591–602.

Meinshausen, M. 2006. "What Does a 2C Target Mean for Greenhouse Gas Concentrations? A Brief Analysis Based on Multi-Gas Emission Pathways and Several Climate Sensitivity Uncertainty Estimates." In *Avoiding Dangerous Climate Change*, ed. H. J. Schellnhuber, W. Cramer, N. Nakicenovic, T. Wigley, and G. Yohe, 253–279. Cambridge: Cambridge University Press.

Meyer, A. 2000. *Contraction & Convergence—The Global Solution to Climate Change*. Devon: Green Books.

Nakicenovic, N., O. Davidson, G. Davis, A. Grubler, T. Kram, E. Lebre La Rovere, B. Metz, T. Morita, W. Pepper, A. Sankovski, P. Shukla, R. Swart, R. Watson, and Z. Dadi. 2000. *IPCC Special Report on Emission Scenarios*. Cambridge: Cambridge University Press.

National Bureau of Statistics China. 2006. *Transport, Post and Telecommunication Services*, 660. China Statistical Yearbook. National Bureau of Statistics. China: China Statistics Press.

Ouzky, M. 2007. "Include Aircraft in Emission Trading Scheme by 2010, Says Environment Committee." http://www.europarl.europa.eu/news/expert/infopress_page/064-11020-275-10-40-911-20071001IPR10999-02-10-2007-2007-false/default_da.htm (accessed October 19, 2007).

Penner, J. E., D. G. Lister, D. J. Griggs, J. Dokken, and M. McFarland, eds. 1999. *Aviation and the Global Atmosphere: A Special Report of IPCC Working Groups I and III*. Cambridge: Cambridge University Press.

PIU. 2002. *The Energy Review*. Performance and Innovation Unit, Cabinet Office, London.

Ponater, M., S. Marquart, K. Strom, and R. Sausen. 2003. *On the Potential of the Cryoplane Technology to Reduce Aircraft Climate Impact*. Germany: AAC.

Pongas, E. 2000. *Air Transport Passenger Traffic 1993–1997: Statistics in Focus*. Eurostat, Brussels.

Raupach, M. R., G. Marland, P. Ciais, C. Le Quere, J. G. Canadell, G. Klepper, and C. B. Field. 2007. "Global and Regional Drivers of Accelerating CO2 Emissions." *PNAS* 104 (24): 10288–10293.

RCEP. 2000. *Energy—The Changing Climate, 22nd Report, CM 4749*. London: The Stationery Office.

———. 2002. *The Environmental Effects of Civil Aircraft in Flight*. Special report of the Royal Commission on Environmental Pollution. Royal Commission on Environmental Pollution, London.

Robinson, J. B. 1982. "Energy Backcasting: A Proposed Method of Policy Analysis." *Energy Policy* 10 (4): 337–344.

Rolls-Royce. 2007. "Market Outlook 2007: Forecast 2007–2026." http://www.rolls-royce.com/civil_aerospace/overview/market/outlook/downloads/outlook06-09-07.pdf.

Sausen, R., I. Isaksen, V. Grewe, D. Hauglustaine, D. S. Lee, G. Myhre, M. O. Kohler, G. Pitari, U. Schumann, F. Stordal, and C. Zerefos. 2005. "Aviation Radiative Forcing in 2000: An Update on IPCC (1999)." *Meteorologische Zeitschrift* 14 (4): 555–561.

Saynor, B., A. Bauen, and M. Leach. 2003. *The Potential for Renewable Energy Sources in Aviation P. F. Report*. London: Imperial College Centre for Energy Policy and Technology.

Stern, N. 2006. *Stern Review on the Economics of Climate Change. Her Majesty's Treasury*. Cambridge: Cambridge University Press.

Tronet, V. 2002. *Air Transport in the Candidate Countries 1995–2000: Statistics in Focus; Transport*. Eurostat, Brussels.
———. 2004. *Passenger Air Transport 2000—2001: Statistics in Focus*. Eurostat Brussels.
UNFCCC. 1997. *The Kyoto Protocol*. United Nations Framework Convention on Climate Change, Kyoto.
———. 2006. "National Inventory Submissions 2006, UNFCCC." http://unfccc.int/national_reports/annex_i_ghg_inventories/national_inventories_submissions/items/3734.php.
Upham, P. J. 2003. "Climate Change and Planning and Consultation for The UK Aviation White Paper." *Journal of Environmental Planning and Management* 46 (6): 911–918.
Warren, R. 2006. "Impacts of Global Climate Change at Different Annual Mean Global Temperature Increases," 93–131. In *Avoiding Dangerous Climate Change*, ed. H. J. Schellnhuber, W. Cramer, N. Nakicenovic, T. Wigley, and G. Yohe. Cambridge: Cambridge University Press.
Williams, V., and R. B. Noland. 2005. "Variability of Contrail Formation Conditions and the Implications for Policies to Reduce the Climate Impacts of Aviation." *Transportation Research Part D: Transport and Environment* 10 (4): 269–280.
Williams, V., R. B. Noland, and R. Toumi. 2003. "Air Transport Cruise Altitude Restrictions to Minimize Contrail Formation." *Climate Policy* 3:207–219.
Windischbauer, F., and J. Richardson. 2005. "Restrostrategy: Is There Another Chance for Lighter-Than-Air Vehicles." *Foresight* 7 (2): 54–65.
Wit, R. C. N., B. H. Boon, A. van Velzen, A. Cames, O. Deuber, and D. S. Lee. 2005. *Giving Wings to Emissions Trading—Inclusion of Aviation under the European Trading System (ETS): Design and Impacts*. CE Delft, Brussels.
Wit, R. C. N., J. W. M. Dings, P. Mendes de Leon, L. Thwaites, P. Peeters, D. Greenwood, and D. R. 2002. *Economic Incentives to Mitigate Greenhouse Gas Emissions from Air Transport in Europe*. CE Delft, Brussels.
World Bank. 2006. *World Development Indicators*. Washington, D.C.: The World Bank Group.

Index